MEMOIRS
of the
American Mathematical Society
ISSN 0065-9266 (print)
ISSN 1947-6221 (online)

Number 1562

AF854879

Mapping Analytic Surgery to Homology, Higher Rho Numbers and Metrics of Positive Scalar Curvature

Paolo Piazza
Thomas Schick
Vito Felice Zenobi

May 2025 • Volume 309 • Number 1562 (third of 6 numbers)

AMERICAN MATHEMATICAL SOCIETY

Library of Congress Cataloging-in-Publication Data

Cataloging-in-Publication Data has been applied for by the AMS.
See http://www.loc.gov/publish/cip/.
DOI: https://doi.org/10.1090/memo/1562

Memoirs of the American Mathematical Society

This journal is devoted entirely to research in pure and applied mathematics.

Subscription information. Beginning in 2024, Memoirs will be published monthly through 2026. Memoirs is also accessible from www.ams.org/journals. The 2025 subscription begins with volume 305 and consists of twelve mailings, each containing one or more numbers. Individual subscription prices for 2025 are as follows. For electronic only: US$1044. For paper delivery: US$1197. Add US$65 for delivery within the United States; US$200 for surface delivery outside the United States. Upon request, subscribers to paper delivery of this journal are also entitled to receive electronic delivery. For information on institutional pricing, please visit `https://www.ams.org/publications/journals/subscriberinfo`. Subscription renewals are subject to late fees. See `www.ams.org/journal-faq` for more journal subscription information. Each number may be ordered separately; please specify number when ordering an individual number.

Back number information. For back issues see `www.ams.org/backvols`.

Subscriptions and orders should be addressed to the American Mathematical Society, P. O. Box 845904, Boston, MA 02284-5904 USA. *All orders must be accompanied by payment.* Other correspondence should be addressed to 201 Charles Street, Providence, RI 02904-2213 USA.

Copying and reprinting. Individual readers of this publication, and nonprofit libraries acting for them, are permitted to make fair use of the material, such as to copy select pages for use in teaching or research. Permission is granted to quote brief passages from this publication in reviews, provided the customary acknowledgment of the source is given.

Republication, systematic copying, or multiple reproduction of any material in this publication is permitted only under license from the American Mathematical Society. Requests for permission to reuse portions of AMS publication content are handled by the Copyright Clearance Center. For more information, please visit `www.ams.org/publications/pubpermissions`.

Send requests for translation rights and licensed reprints to `reprint-permission@ams.org`.

Excluded from these provisions is material for which the author holds copyright. In such cases, requests for permission to reuse or reprint material should be addressed directly to the author(s). Copyright ownership is indicated on the copyright page, or on the lower right-hand corner of the first page of each article within proceedings volumes.

Memoirs of the American Mathematical Society (ISSN 0065-9266 (print); 1947-6221 (online)) is published bimonthly (each volume consisting usually of more than one number) by the American Mathematical Society at 201 Charles Street, Providence, RI 02904-2213 USA. Periodicals postage paid at Providence, RI. Postmaster: Send address changes to Memoirs, American Mathematical Society, 201 Charles Street, Providence, RI 02904-2213 USA.

© 2025 by the American Mathematical Society. All rights reserved.

This publication is indexed in *Mathematical Reviews®*, *Zentralblatt MATH*, *Science Citation Index®*, *Science Citation Index™-Expanded*, *ISI Alerting Services℠*, *SciSearch®*, *Research Alert®*, *CompuMath Citation Index®*, *Current Contents®/Physical, Chemical & Earth Sciences.*
This publication is archived in *Portico* and *CLOCKSS.*
Printed in the United States of America.

♾ The paper used in this book is acid-free and falls within the guidelines established to ensure permanence and durability.
Visit the AMS home page at `https://www.ams.org/`

10 9 8 7 6 5 4 3 2 1   30 29 28 27 26 25

Contents

Abstract

Let Γ be a finitely generated discrete group and let $\widetilde{M}$ be a Galois Γ-covering of a smooth compact manifold M. Let $u\colon M \to B\Gamma$ be the associated classifying map. Finally, let $\mathrm{S}^\Gamma_*(\widetilde{M})$ be the analytic structure group, a K-theory group appearing in the Higson-Roe analytic surgery sequence $\cdots \to \mathrm{S}^\Gamma_*(\widetilde{M}) \to K_*(M) \to K_*(C^*_{red}\Gamma) \to \cdots$. Under suitable assumptions on the group Γ we construct two pairings, first between $\mathrm{S}^\Gamma_*(\widetilde{M})$ and the delocalized part of the cyclic cohomology of $\mathbb{C}\Gamma$ and secondly between $\mathrm{S}^\Gamma_*(\widetilde{M})$ and the relative cohomology $H^{*-1}(M \to B\Gamma)$. Both are compatible with known pairings associated with the other terms in the Higson-Roe sequence. In particular, we define higher rho numbers associated to the rho class $\rho(\widetilde{D}) \in \mathrm{S}^\Gamma_*(\widetilde{M})$ of an invertible Γ-equivariant Dirac type operator on $\widetilde{M}$. This applies, for example, to the rho class $\rho(g)$ of a positive scalar curvature metric g on M, if M is spin. Regarding the first pairing, we establish in fact a more general result. Indeed, for an arbitrary discrete group Γ we prove that it is possible to map the whole Higson-Roe analytic surgery sequence to the long exact sequence in even/odd-graded non-commutative de Rham homology, $\cdots \xrightarrow{j_*} H_{*-1}(\mathcal{A}\Gamma) \xrightarrow{\iota} H^{del}_{*-1}(\mathcal{A}\Gamma) \xrightarrow{\delta} H^e_*(\mathcal{A}\Gamma) \xrightarrow{j_*} \cdots$ with $\mathcal{A}\Gamma$ any dense holomorphically closed subalgebra of $C^*_{red}\Gamma$. Here, $H^{del}_*(\mathcal{A}\Gamma)$ and $H^e_*(\mathcal{A}\Gamma)$ denote versions of the delocalized homology and the homology localized at the identity element, respectively. In particular, we prove that there exists a group homomorphism $\mathrm{S}^\Gamma_*(\widetilde{M}) \to H^{del}_{*-1}(\mathcal{A}\Gamma)$. We then establish that under additional assumptions on Γ there exists a pairing between the delocalized part of the cyclic cohomology of $\mathbb{C}\Gamma$ and $H^{del}_*(\mathcal{A}\Gamma)$.

We give a detailed study of the action of the diffeomorphism group on the analytic surgery sequence and on its homological counterpart, proving in particular precise formulae for the behaviour of the higher rho numbers under the action of the diffeomorphism group. We use these formulae in order to establish new lower bounds on the size of the moduli space of metrics of positive scalar curvature.

Received by the editor February 23, 2021, and, in revised form, October 27, 2022.

Article electronically published on April 22, 2025.

DOI: https://doi.org/10.1090/memo/1562

2020 *Mathematics Subject Classification.* Primary 46L80; Secondary 19D55, 19K56, 53C21, 53C27.

Key words and phrases. Secondary index, rho invariants, Chern character, non-commutative geometry, positive scalar curvature.

©2025 American Mathematical Society

CHAPTER 1

Introduction

Let (M, g) be a connected smooth compact Riemannian manifold without boundary. We denote by Γ the fundamental group $\pi_1(M)$ and by $\widetilde{M}$ the associated universal cover. For the time being we assume that M is even dimensional. Let $E = E^+ \oplus E^- \to M$ be a $\mathbb{Z}_2$-graded bundle of Clifford modules and

$$D = \begin{pmatrix} 0 & D^- \\ D^+ & 0 \end{pmatrix} : C^\infty(M, E) \to C^\infty(M, E)$$

a self-adjoint generalized Dirac operator. Relevant examples are given by the signature operator if we assume that M is orientable, and the spin Dirac operator if, in addition, M is spin. We can lift all the data defining D to $\widetilde{M}$ and obtain in this way a Γ-equivariant Dirac operator $\widetilde{D}$.

One of the most basic invariant that can be associated to D is the Fredholm index of D^+:

$$\begin{aligned} \operatorname{ind}(D^+) &= \dim\ker D^+ - \dim\operatorname{coker} D^+ \\ &= \dim\ker D^+ - \dim\ker D^- \\ &= \operatorname{Tr}\Pi^+ - \operatorname{Tr}\Pi^- \end{aligned}$$

with $\Pi^\pm$ the orthogonal projections onto $\ker D^\pm$. The Atiyah-Singer index theorem, one of the milestones of modern mathematics, provides a geometric formula for this analytic invariant. One very useful expression for the index of D^+ is given in terms of the remainders associated to a parametrix Q: if

$$D^+Q = \operatorname{Id} - S_- \quad QD^+ = \operatorname{Id} - S_+ \quad \text{with} \quad S_\pm \in \Psi^{-\infty}(M, E^\pm),$$

where $\Psi^{-\infty}(M, E^\pm)$ denotes the algebra of smoothing pseudodifferential operators, then

$$\operatorname{ind}(D^+) = \operatorname{Tr} S_+^N - \operatorname{Tr} S_-^N \quad \forall N \geq 1. \tag{1.1}$$

We can also consider the Γ-index associated to $\widetilde{D}$, as in [**1**]. In order to define it we consider the von Neumann algebra $\mathcal{A}_\Gamma$ of all bounded Γ-equivariant operators on $L^2(\widetilde{M}, \widetilde{E})$; the orthogonal projections onto $\operatorname{Ker}\widetilde{D}^\pm$, denoted $\Pi_\Gamma^\pm$, are elements in $\mathcal{A}_\Gamma$. The von Neumann algebra $\mathcal{A}_\Gamma$ comes equipped with a canonical trace Tr_Γ and the orthogonal projections $\Pi_\Gamma^\pm$ are trace class with respect to this von Neumann trace. It thus makes sense to define

$$\operatorname{ind}_\Gamma(\widetilde{D}^+) := \operatorname{Tr}_\Gamma \Pi_\Gamma^+ - \operatorname{Tr}\Pi_\Gamma^-$$

and a fundamental result of Atiyah asserts that

$$\operatorname{ind}(D^+) = \operatorname{ind}_\Gamma(\widetilde{D}^+).$$

A major step in the proof of this equality consists in showing that we can express the Γ-index of $\widetilde{D}^+$ in terms of the remainders associated to a suitable Γ-equivariant parametrix $\widetilde{Q}$ of Γ-compact support, where a Γ-equivariant pseudodifferential operator is defined to be of Γ-compact support if the support of its Schwartz kernel has compact image when projected in $(\widetilde{M} \times \widetilde{M})/\Gamma$, where we divide by the diagonal action. We denote the smoothing Γ-equivariant operators of Γ-compact support by $\Psi^{-\infty}_{\Gamma,c}(\widetilde{M}, \widetilde{E})$ and recall that if $\widetilde{S} \in \Psi^{-\infty}_{\Gamma,c}(\widetilde{M}, \widetilde{E})$ then $\widetilde{S}$ has finite Γ-trace computed by

$$\operatorname{Tr}_\Gamma(\widetilde{S}) = \int_{\mathcal{F}} \operatorname{tr}_p K_{\widetilde{S}}(p,p) \operatorname{dvol}$$

with $K_{\widetilde{S}}$ the smooth integral kernel of $\widetilde{S}$ and $\mathcal{F}$ a fundamental domain for the action of Γ on $\widetilde{M}$.

Going back to the Γ-index, it is shown by Atiyah in [**1**] that for every parametrix $\widetilde{Q}$ such that

$$\widetilde{D}^+\widetilde{Q} = \operatorname{Id} - \widetilde{S}_- \quad \widetilde{Q}\widetilde{D}^+ = \operatorname{Id} - \widetilde{S}_+ \quad \text{with} \quad \widetilde{S}_\pm \in \Psi^{-\infty}_{\Gamma,c}(\widetilde{M}, \widetilde{E}^\pm)$$

one can compute the Γ-index by

$$\operatorname{ind}_\Gamma(\widetilde{D}^+) = \operatorname{Tr}_\Gamma \widetilde{S}_+^N - \operatorname{Tr}_\Gamma \widetilde{S}_-^N \quad \forall N \geq 1. \tag{1.2}$$

We can recover these invariants using K-theory. Let Q be a parametrix for D with remainders $S_\pm \in \Psi^{-\infty}(M, E^\pm)$. Consider the 2×2 matrix

$$P := \begin{pmatrix} S_+^2 & S_+(I + S_+)Q \\ S_- D^+ & I - S_-^2 \end{pmatrix}.$$

This is an idempotent matrix over the unitalization of $\Psi^{-\infty}(M, E)$; also

$$e_1 := \begin{pmatrix} 0 & 0 \\ 0 & 1_{E^-} \end{pmatrix}$$

is an idempotent matrix in the unitalization and we define the index class associated to D as

$$\operatorname{Ind}(D) := [P] - [e_1] \in K_0(\Psi^{-\infty}(M, E)). \tag{1.3}$$

This definition comes from the short exact sequence of algebras

$$0 \to \Psi^{-\infty}(M, E) \to \Psi^0(M, E) \to \Psi^0(M, E)/\Psi^{-\infty}(M, E) \to 0$$

and the associated long exact sequence in K-theory: the K-theory class (1.3) is nothing but the boundary map applied to the K_1-class given by the bounded transform of D (which is indeed invertible in the quotient). See [**20**] for details. The functional analytic trace Tr defines a cyclic cocycle of degree 0 on the algebra $\Psi^{-\infty}(M, E)$ and from the pairing of K-theory with cyclic cohomology we have a well-defined number $\langle \operatorname{Ind}(D), [\operatorname{Tr}] \rangle$; from the expression of $\operatorname{Ind}(D)$ and formula (1.1) we deduce

$$\langle \operatorname{Ind}(D), [\operatorname{Tr}] \rangle = \operatorname{Tr} S_+^2 - \operatorname{Tr} S_-^2 = \operatorname{ind}(D^+).$$

Similarly, we have an index class

$$\operatorname{Ind}_{\Gamma,c}(\widetilde{D}^+) := [\widetilde{P}] - [e_1] \in K_0(\Psi^{-\infty}_{\Gamma,c}(\widetilde{M}, \widetilde{E}))$$

defined using a parametrix of Γ-compact support for $\widetilde{D}$; the Γ-trace Tr_Γ defines an element in $HC^0(\Psi^{-\infty}_{\Gamma,c}(\widetilde{M}, \widetilde{E}))$ and we get

$$\langle \operatorname{Ind}_{\Gamma,c}(\widetilde{D}), [\operatorname{Tr}_\Gamma] \rangle = \operatorname{Tr} \widetilde{S}_+^2 - \operatorname{Tr} \widetilde{S}_-^2 = \operatorname{ind}_\Gamma(D^+)$$

where we have now used (1.2).

Summarizing: we have recovered the index by pairing the index *class* $\text{Ind}(D) \in K_0(\Psi^{-\infty}(M,E))$ with a 0-degree cyclic cocycle $[\text{Tr}] \in HC^0(\Psi^{-\infty}(M,E))$ and we have recovered the Γ-index by pairing the index class $\text{Ind}_{\Gamma,c}(\widetilde{D}) \in K_0(\Psi^{-\infty}_{\Gamma,c}(\widetilde{M},\widetilde{E}))$ with the 0-degree cyclic cocycle $[\text{Tr}_\Gamma] \in HC^0(\Psi^{-\infty}_{\Gamma,c}(\widetilde{M},\widetilde{E}))$.

We are led naturally to define *higher indices* by pairing the index class

$$\text{Ind}_{\Gamma,c}(\widetilde{D}) \in K_0(\Psi^{-\infty}_{\Gamma,c}(\widetilde{M},\widetilde{E}))$$

with *higher cyclic cocycles* for the algebra $\Psi^{-\infty}_{\Gamma,c}(\widetilde{M},\widetilde{E})$, that is with elements in $HC^{\text{even}}(\Psi^{-\infty}_{\Gamma,c}(\widetilde{M},\widetilde{E}))$. One reason to pass directly to the index class associated to $\widetilde{D}$ and disregard the index class associated to D is that the K-theory and the cyclic cohomology of $\Psi^{-\infty}_{\Gamma,c}(\widetilde{M},\widetilde{E})$ are much richer than the corresponding groups for $\Psi^{-\infty}(M,E)$. For example, elements in $H^k(\Gamma,\mathbb{C})$ define in a natural way elements in $HC^k(\mathbb{C}\Gamma)$ which in turn define elements in $HC^k(\Psi^{-\infty}_{\Gamma,c}(\widetilde{M},\widetilde{E}))$. The Connes-Moscovici higher index theorem [**20**] provides a geometric formula for these higher indices; this is of great interest both intrinsically and for geometric applications.

Another important reason to consider operators on $\widetilde{M}$ is the following: for geometric applications, one is eventually interested in the image of these index classes in the K-theory of the C^*-closures of these algebra. Now, the C^*-closure of $\Psi^{-\infty}(M,E)$ in $\mathcal{B}(L^2(M,E))$ is equal to the compact operators, whose K-theory is well known, equal to $\mathbb{Z}$ in even degree and equal to 0 in odd degree; this means that the *C^*-index class* on M retains essentially the same information as the numeric index. The K-theory of the C^*-closure of $\Psi^{-\infty}_{\Gamma,c}(\widetilde{M},\widetilde{E})$ in $\mathcal{B}(L^2(\widetilde{M},\widetilde{E}))$ is, on the other hand, much richer, as it is isomorphic to the K-theory of $C^*_{red}\Gamma$, the reduced C^*-algebra of Γ. The C^*-closure of $\Psi^{-\infty}_{\Gamma,c}(\widetilde{M},\widetilde{E})$ in $\mathcal{B}(L^2(\widetilde{M},\widetilde{E}))$ is called the *Roe algebra* of $\widetilde{M}$ and it is denoted by $C^*(\widetilde{M},\widetilde{E})^\Gamma$; we denote the index class in the Roe algebra of $\widetilde{M}$ by $\text{Ind}_\Gamma(\widetilde{D})$. Thus

$$\text{Ind}_\Gamma(\widetilde{D}) \in K_0(C^*(\widetilde{M},\widetilde{E})^\Gamma) = K_0(C^*_{red}\Gamma).$$

At this point, it is important to point out that the higher indices we have defined are *always* defined for the compactly supported index class $\text{Ind}_{\Gamma,c}(\widetilde{D})$ and computable through the Connes-Moscovici formula. However, the most relevant geometric properties of the index cannot be derived from this index class but only from the higher C^*-indices, that is higher indices associated to $\text{Ind}_\Gamma(\widetilde{D}) \in K_0(C^*(\widetilde{M},\widetilde{E})^\Gamma)$. This refers in particular to vanishing results: the higher C^*-index of the Dirac operator of a spin manifold vanishes if the scalar curvature is positive. Similarly the difference of the higher C^*-indices of the signature operators of two homotopy equivalent manifolds vanish. The corresponding statements in general are not known to be true for the compactly supported index class $\text{Ind}_{\Gamma,c}(\widetilde{D})$. The vanishing results follow from the fact that positivity implies invertibility in C^*-algebras (but this is not true for general $*$-algebras).

To obtain numerical C^*-algebraic indices, we need an additional argument (and possibly additional hypothesis) ensuring that the relevant pairing extends from the algebra $\Psi^{-\infty}_{\Gamma,c}(\widetilde{M},\widetilde{E})$ to its C^*-closure, that is, to the Roe algebra $C^*(\widetilde{M},\widetilde{E})^\Gamma$. Equivalently, if we consider the index class as an element in $K_0(C^*_{red}\Gamma)$ we shall need additional arguments and possibly additional hypothesis ensuring that we can *extend* cyclic cocycles for the algebra $\mathbb{C}\Gamma$ to the C^*-algebra $C^*_{red}\Gamma$. We shall come back to this crucial point later in this introduction.

To give an example of this circle of ideas let us consider the flat n-torus T^n, with $n = 2k$. We know that on a spin manifold the index of the spin Dirac operator, which by the Atiyah-Singer index theorem is the $\widehat{A}$-genus, is an obstruction to the existence of a metric of positive scalar curvature. The $\widehat{A}$-genus of the flat torus vanishes (because its tangent bundle is trivial) and so we find ourselves without valuable information as to whether T^n admits or not a metrics of positive scalar curvature. Now $\pi_1(T^n) = \mathbb{Z}^n$, $C^*_{red}\mathbb{Z}^n = C(\widehat{T}^n)$, with $\widehat{T}^n$ the dual torus and $H^*(\mathbb{Z}^n, \mathbb{C}) \simeq \Lambda^*\mathbb{C}^n$. For this C^*-algebra it is indeed possible to extend the cyclic cocycles coming from $H^*(\mathbb{Z}^n, \mathbb{C})$ and a direct application of the Connes-Moscovici higher index formula shows that the higher index of the spin Dirac operator associated to the generator of $H^n(T^n)$ is equal to a non-zero multiple of the volume of T^n and thus different from zero. This implies of course that the index class $\mathrm{Ind}_{\mathbb{Z}^n}(\widetilde{D}^+)$ is different from 0 and we conclude that T^n does not admit a metric of positive scalar curvature; this conclusion has been made possible by passing from the lower index $\mathrm{ind}(D^+)$ to a higher index.

So far, we have restricted ourselves to the even dimensional case, but it is also possible to introduce the higher index class in the odd dimensional case, obtaining

$$\mathrm{Ind}_{\Gamma,c}(\widetilde{D}) \in K_1(\Psi^{-\infty}_{\Gamma,c}(\widetilde{M}, \widetilde{E})).$$

Higher numerical indices are then obtained by pairing with $HC^{\mathrm{odd}}(\Psi^{-\infty}_{\Gamma,c}(\widetilde{M}, \widetilde{E}))$. As in the even dimensional case, it is then even more important to consider

$$\mathrm{Ind}_{\Gamma}(\widetilde{D}) \in K_1(C^*(\widetilde{M}, \widetilde{E})^{\Gamma})$$

and the possibility of extracting numerical invariants out of it which is indeed granted, as for even dimensions, under additional hypothesis on Γ. We shall explain this momentarily.

Despite their many interesting applications, the index invariants we have introduced will be inadequate whenever we have a geometric situation in which the index classes vanish. This is the case, for example, if D is the spin Dirac operator associated to a metric of positive scalar curvature or if D is the signature operator on the disjoint union of two homotopy equivalent manifolds with opposite orientation. In geometric questions involving, for example, the diffeomorphism type of homotopy equivalent manifolds or the moduli space of metrics of positive scalar curvature $\mathcal{R}^+(M)/\mathrm{Diffeo}(M)$, one is led to consider *secondary* invariants, the rho invariants associated to D and $\widetilde{D}$. In their basic version, analogous to the numeric indices $\mathrm{ind}(D^+)$ and $\mathrm{ind}_{\Gamma}(\widetilde{D}^+)$, these are defined for a self-adjoint Dirac operator and we list the three most important versions:

- the Atiyah-Patodi-Singer rho invariant $\rho_{\alpha-\beta}(D)$ associated to two unitary representations $\alpha, \beta\colon \pi_1(M) \to U(\ell)$ as introduced in [**2**];
- the Cheeger-Gromov L^2-rho invariant $\rho_{(2)}(\widetilde{D})$ of [**17**];
- the delocalized eta invariant of Lott $\eta_{\langle x\rangle}(\widetilde{D})$ associated, for the time being, to a finite conjugacy class $\langle x\rangle$ in $\Gamma = \pi_1(M)$ and for $\widetilde{D}$ assumed here to be invertible, as constructed in [**53**].

Results of Botvinnik-Gilkey [**12**], Chang-Weinberger [**16**], Piazza-Schick [**64**] and subsequent work show that these invariants are extremely useful in the geometric questions we have presented above, that is, the problem of distinguishing the diffeomorphism type of two homotopically equivalent manifolds or the problem of distinguishing components of the moduli space of metrics of positive scalar curvature, i.e. elements of $\pi_0(\mathcal{R}^+(M)/\operatorname{Diffeo}(M))$.

We now come to a fundamental question, which is the analogue of the passage from the numeric *lower* index invariants $\operatorname{ind}(D^+)$ and $\operatorname{ind}_\Gamma(\widetilde{D}^+)$ to the *higher* indices associated to the index classes

$$\operatorname{Ind}_{\Gamma,c}(\widetilde{D}) \in K_*(\Psi^{-\infty}_{\Gamma,c}(\widetilde{M},\widetilde{E})) \quad \text{and} \quad \operatorname{Ind}_\Gamma(\widetilde{D}) \in K_*(C^*(\widetilde{M},\widetilde{E})^\Gamma)$$

and the cyclic cocycles for the algebra $\mathbb{C}\Gamma$:

Question: *can we define a rho class in the K-theory of a suitable C^*-algebra and produce the lower rho invariants introduced above as well as new, higher numeric rho invariants by pairing this rho class with suitable cyclic cocycles?*

Notice that these are really two distinct questions. We ask whether it is possible to

(1) produce a rho class in the K-theory of a suitable C^*-algebra;
(2) pair this rho class with suitable cyclic cocycles so as to define and compute higher rho invariants.

The answer to the first question is positive and is fundamental work of Higson and Roe. Let us expunge the vector bundle E from the notation and let us denote the Roe C^*-algebra by $C^*(\widetilde{M})^\Gamma$. In their seminal paper [**30**] Higson and Roe construct a long exact sequence of K-theory groups

(1.4)
$$\cdots \to K_{*+1}(C^*(\widetilde{M})^\Gamma) \to K_{*+1}(D^*(\widetilde{M})^\Gamma) \to K_{*+1}(D^*(\widetilde{M})^\Gamma/C^*(\widetilde{M})^\Gamma) \to \cdots$$

We call this sequence the *Higson-Roe exact sequence* or the *Higson-Roe analytic surgery sequence*. We shall explain in a moment why the word surgery is employed in this context.

This sequence organizes and captures in a very elegant way the interplay of primary versus secondary index information of Dirac operators defined on M and $\widetilde{M}$. As already explained, first of all we have a canonical isomorphism

$$K_*(C^*(\widetilde{M})^\Gamma) \cong K_*(C^*_{red}\Gamma)\,.$$

Next we have that

$$K_{*+1}(D^*(\widetilde{M})^\Gamma/C^*(\widetilde{M})^\Gamma) \cong K_*(M)\,, \tag{1.5}$$

the K-homology of M. Here, K-homology is the homology theory dual to topological K-theory, and the isomorphism (1.5) is called *Paschke duality.* Moreover, under these isomorphisms the connecting homomorphism of the long exact sequence (1.4), namely

$$K_{*+1}(D^*(\widetilde{M})^\Gamma/C^*(\widetilde{M})^\Gamma) \to K_*(C^*(\widetilde{M})^\Gamma),$$

becomes the Baum-Connes assembly map

$$K_*(M) = K^\Gamma_*(\widetilde{M}) \xrightarrow{\mu} K_*(C^*_{red}\Gamma).$$

Setting

$$\mathrm{S}^\Gamma_*(\widetilde{M}) := K_{*+1}(D^*(\widetilde{M})^\Gamma),$$

we can rewrite the Higson-Roe surgery sequence as

$$\cdots \to K_{*+1}(C^*_{red}\Gamma) \xrightarrow{s} \mathrm{S}^\Gamma_*(\widetilde{M}) \xrightarrow{c} K_*(M) \xrightarrow{\mu} K_*(C^*_{red}\Gamma) \to \cdots$$

We know that the group $K_*(C^*_{red}\Gamma)$ is the home of index invariants associated to Γ-equivariant geometric operators $\widetilde{M}$.

The group $\mathrm{S}^\Gamma_*(\widetilde{M})$ is called the *analytic structure group* by Higson and Roe and it is the home of the K-theoretic secondary invariants we are looking for. In particular, if M is a spin manifold with a positive scalar curvature metric g, one obtains a secondary invariant $\rho(g) \in \mathrm{S}^\Gamma_{\dim M}(\widetilde{M})$, called the rho class of g, which captures information about the positive scalar curvature metric g, and allows often to distinguish different bordism classes of such metrics. Similarly, if $N \xrightarrow{f} M$ is an oriented homotopy equivalence, then we have a rho class $\rho(f) \in \mathrm{S}^\Gamma_{\dim M}(\widetilde{M})$ and this class can distinguish triples $N' \xrightarrow{f'} M$ and $N \xrightarrow{f} M$ that are not h-cobordant or, put it differently, triples that define distinct elements in the manifold structure set $\mathcal{S}(M)$.

More generally, the rho class can be defined for any L^2-invertible Γ-equivariant Dirac operator $\widetilde{D}$ on $\widetilde{M}$, $\rho(\widetilde{D}) \in \mathrm{S}^\Gamma_{\dim M}(\widetilde{M})$; in fact, we can allow for invertible smoothing perturbations of such operators, see [**66**].

The construction of the K-theoretic rho classes follows the following abstract principle, which is explained in more detail and with more examples in [**70**].

(1) The geometry gives as a fundamental class $F \in K_*(M)$. In the example of a spin manifold, $F = [D]$ is indeed the K-homology fundamental class of the Dirac operator D (or, formulated in topological terms, of the spin structure).
(2) The map μ is an index map. It maps the fundamental class F to the K-theoretic index class $\mathrm{Ind}_\Gamma(\widetilde{D}) \in K_*(C^*_{red}(\Gamma))$. In the example of a spin manifold, this is indeed the higher index class of the Dirac operator. In case of an oriented manifold, we consider the index class of the signature operator.
(3) We have the following observation: by the exactness of the Higson-Roe sequence, $\mathrm{Ind}_\Gamma(\widetilde{D}) = 0$ if and only if there are classes $a \in \mathrm{S}^\Gamma_*(\widetilde{M})$ with $c(a) = F$. However, the vanishing of $\mathrm{Ind}_\Gamma(\widetilde{D})$ by itself does not yet determine a specific lift a of the fundamental class.
(4) What is needed, additionally, is a *geometric reason* for the vanishing of the index. In favourable situations, as we will describe them now, the geometry not only will imply the vanishing of the index class, but will give rise to a specific lift $\rho \in \mathrm{S}^\Gamma_*(\widetilde{M})$ of F (i.e. $c(\rho) = F$).
(5) This arises as follows: in our definition of the Higson-Roe exact sequence, $K_*(M)$ is defined as the K-theory of the quotient $D^*(\widetilde{M})^\Gamma / C^*(\widetilde{M})^\Gamma$ of two C^*-algebras. The fundamental class is explicitly given by an element F in this quotient algebra, which is indeed obtained from the underlying geometric operator (of Dirac type). Note that, to define a K-theory class, the element F has to be invertible (to define a K_1-class) or idempotent (to define a K_0-class), and this property for general geometric situation is only satisfied in $D^*(\widetilde{M})^\Gamma / C^*(\widetilde{M})^\Gamma$.

   However, if one has additional special geometry (in particular, positive scalar curvature in the case of the Dirac operator), then the operator

F already is invertible/an idempotent in the algebra $D^*(\widetilde{M})^\Gamma$ and therefore defines a class in its K-theory. Slightly more generally, the special geometry (in particular, a homotopy equivalence for the signature operator on the disjoint union of the two homotopy equivalent manifolds) gives a canonical perturbation of F which is invertible/an idempotent in $D^*(\widetilde{M})^\Gamma$.

In any case, this invertible/idempotent now defines a class in the K-theory of $D^*(\widetilde{M})^\Gamma$, i.e. in $\mathrm{S}^\Gamma_*(\widetilde{M})$. By definition, this is the rho class we want to construct. By its very construction, it is a lift of the fundamental class F and it depends on and therefore potentially contains information about the geometric reason for the vanishing of the index class.

As already mentioned, there are two main domains where the Higson-Roe exact sequence is used: signature invariants of orientable manifolds, as part of surgery theory to classify smooth manifolds, and classifications of metrics of positive scalar curvature. The first domain is organized in the celebrated *surgery exact sequence* of Browder, Novikov, Sullivan and Wall, and the connection between topology and analysis is made systematic by the construction of a map of exact sequences, from the surgery exact sequence in topology to the Higson-Roe analytic surgery sequence. This explains the word surgery in the exact sequence of Higson and Roe. This is carried out originally in [**30**–**32**] using analytic homological algebra. A more direct construction of this map, based on certain higher index theorems, is given in [**66**].

This second approach is in fact inspired by the precursor work [**65**], where the Stolz exact sequence for positive scalar curvature metrics (compare [**65**, Proposition 1.27]) is mapped to the Higson-Roe exact sequence. The Stolz exact sequence is the analogue, in the world of positive scalar curvature metrics, of the surgery exact sequence in differential topology; it organizes questions about existence and (bordism) classification of Riemannian metrics of positive scalar curvature in a long exact sequence of groups, but where the main terms are somewhat hard to analyze directly.

The results of Piazza and Schick about the Stolz surgery sequence [**65**] were established in odd dimensions. Subsequently, Xie and Yu gave a different treatment of these results, valid in all dimensions and based on Yu's localization algebras (see [**85**], [**82**]). The even dimensional case of the results of Piazza-Schick was later treated by Zenobi in [**88**], using suspension. Yet different treatments of these results were given by Zeidler in [**86**] and by Zenobi in [**90**].

Both for the surgery sequence in topology and for the Stolz sequence for positive scalar curvature metrics, the map to the Higson-Roe exact sequence allows to apply all the tools from C^*-algebraic index theory and C^*-algebra K-theory to obtain information about the geometrically defined sequences. In this toolbox we find, in particular, homological methods.

This has a long history for primary index invariants, as we have seen earlier in this introduction, thanks in particular to the seminal work of Connes and Moscovici on the higher index formula and its applications to the Novikov conjecture [**20**]. As already hinted to, this geometric application is made possible only if we are able to *extend* the cyclic cocycles associated to elements in $H^*(\Gamma, \mathbb{C})$ from $\mathbb{C}\Gamma$ to $C^*_{red}\Gamma$, defining in this way higher C^*-indices. This crucial step is achieved by Connes and Moscovici under the additional assumption that Γ is Gromov hyperbolic; the proof of the extendability property for cyclic cocycles defined by group cocycles of Gromov hyperbolic groups is in fact rather subtle.

Lott, inspired by the work of Bismut on the family index theorem, gave a different treatment of the Connes-Moscovici theorem, both for defining the higher indices and for computing them. This approach employs in a crucial way the Chern character homomorphism of Karoubi [**40**], from $K_*(C^*_{red}\Gamma) \equiv K_*(\mathcal{A}\Gamma)$ to the non-commutative de Rham homology groups $H_*(\mathcal{A}\Gamma)$, with $\mathcal{A}\Gamma$ a suitable smooth subalgebra of $C^*_{red}\Gamma$. Here we recall that the non-commutative de Rham homology groups of an algebra are subgroups of its cyclic homology groups. Lott establishes an explicit formula for the Chern character of the index class and this formula is valid without any assumption on Γ. Then, in a second stage, one can construct pairings with group cocycles of Γ by employing a classic result of Burghelea which identifies the group cohomology $H^*(\Gamma, \mathbb{C})$ with a subgroup of the cyclic cohomology of $\mathbb{C}\Gamma$. This way, one obtains numerical invariants for the K-theory classes. For these pairings, one needs assumptions on the group like Connes and Moscovici, for example that Γ is Gromov hyperbolic or of polynomial growth.

We emphasise the following conclusion from the above discussion: by mapping the index class first to non-commutative de Rham homology and *then* pairing the result with cyclic cohomology (this second step under additional assumption on Γ) it has been possible to define *higher indices* for a Γ-equivariant Dirac operator on $\widetilde{M}$. We shall soon make use of this fundamental observation.

Let us go back to secondary invariants. Having explained how Higson and Roe solved in a positive way the existence of a rho class, thus giving a positive answer to the first question raised above, we can now concentrate on the second question, which we reformulate in the following more precise way:

Question: *can one define higher rho numbers associated to the rho class $\rho(\widetilde{D}) \in \mathrm{S}^{\Gamma}_{\dim M}(\widetilde{M})$ of an invertible Γ-equivariant Dirac operator?*

We shall give a positive answer to this question in two ways. First, following Burghelea [**13**], denote by $HC^*(\mathbb{C}\Gamma; \langle x\rangle)$ the factor of the cyclic cohomology of $\mathbb{C}\Gamma$ supported on a conjugacy class $\langle x\rangle$ of Γ. If Γ is Gromov hyperbolic, we prove that there is a pairing

$$(\cdot,\cdot) \,:\, HC^*(\mathbb{C}\Gamma; \langle x\rangle) \times \mathrm{S}^{\Gamma}_*(\widetilde{M}) \to \mathbb{C} \quad \forall x \neq e. \tag{1.6}$$

A construction similar to this pairing is carried out independently in [**18**]. We shall compare more thoroughly the two papers at the end of this introduction.

As a second answer to our basic question we also prove in this paper that under different assumptions on Γ, satisfied for example by groups that are Gromov hyperbolic or of polynomial growth, there is a well defined pairing

$$(\cdot,\cdot)_{\mathrm{rel}} \,:\, H^*(M \xrightarrow{u} B\Gamma) \times \mathrm{S}^{\Gamma}_*(\widetilde{M}) \to \mathbb{C}\,, \tag{1.7}$$

where $H^*(M \xrightarrow{u} B\Gamma)$ denotes the relative singular homology of the map $u\colon M \to B\Gamma$.

We shall now put these two results in context. The first result should be thought of as the analogue of the Connes-Moscovici pairing but done here for the group $\mathrm{S}^{\Gamma}_*(\widetilde{M})$, which is the recipient of secondary invariants, instead of $K_*(C^*_{red}\Gamma)$, which is the recipient of primary invariants.

Following Lott's approach to the Connes-Moscovici higher index theorem, *we shall in fact solve a much more general problem.* Indeed, without any additional assumption on the group Γ we shall map the whole Higson-Roe analytic surgery sequence to a long exact sequence in non-commutative de Rham homology:

$$
\begin{array}{ccccccccc}
\cdots \longrightarrow & K_{*-1}(C^*_{red}\Gamma) & \xrightarrow{s} & \mathrm{S}^\Gamma_*(\widetilde{M}) & \xrightarrow{c} & K^\Gamma_*(\widetilde{M}) & \longrightarrow \cdots \\
& \downarrow \mathrm{Ch}_\Gamma & & \downarrow \mathrm{Ch}^{del}_\Gamma & & \downarrow \mathrm{Ch}^e_\Gamma & \\
\cdots \xrightarrow{j_*} & H_{*-1}(\mathcal{A}\Gamma) & \xrightarrow{\iota} & H^{del}_{*-1}(\mathcal{A}\Gamma) & \xrightarrow{\delta} & H^e_*(\mathcal{A}\Gamma) & \xrightarrow{j_*} \cdots
\end{array}
\tag{1.8}
$$

with $\mathcal{A}\Gamma$ any dense holomorphically closed subalgebra of $C^*_{red}\Gamma$. In a second stage, building on work of Puschnigg and also Meyer, we pair the "delocalized parts" of the cyclic cohomology of $\mathbb{C}\Gamma$, $HC^*_{\mathrm{del}}(\mathbb{C}\Gamma;\langle x\rangle)$, with elements in $H^{del}_*(\mathcal{A}\Gamma)$, the "delocalized part" of the non-commutative de Rham homology of a (suitable) smooth subalgebra $\mathcal{A}\Gamma$ of $C^*_{red}\Gamma$. It is only at this point that we need to make additional assumptions on the group Γ, for example that Γ is Gromov hyperbolic. In this case the smooth subalgebra $\mathcal{A}\Gamma$ is the one constructed by Puschnigg in [**68**]. Thus, under this additional assumption we show that there is a well defined pairing

$$
\langle\cdot,\cdot\rangle : HC^*(\mathbb{C}\Gamma;\langle x\rangle) \times H^{del}_*(\mathcal{A}\Gamma) \to \mathbb{C} \quad \forall e \neq x \in \Gamma \tag{1.9}
$$

and we can therefore define the higher rho number associated to the rho class $\rho(\widetilde{D})$ and a class $\tau \in HC^*_{\mathrm{del}}(\mathbb{C}\Gamma;\langle x\rangle)$ as

$$
\rho_\tau(\widetilde{D}) := \langle \tau, \mathrm{Ch}^{del}(\rho(\widetilde{D}))\rangle. \tag{1.10}
$$

Notice that the existence of the pairing (1.9) has an intrinsic interest that goes beyond the problem of defining higher rho numbers.

As an easy application of our analysis we prove that by pairing the rho class $\rho(\widetilde{D})$ with the delocalized trace associated to a non-trivial conjugacy class $\langle g\rangle$,

$$
\tau_{\langle g\rangle}\Big(\sum_\gamma \alpha_\gamma \gamma\Big) := \sum_{\gamma\in\langle g\rangle} \alpha_\gamma\,,
$$

we obtain precisely the delocalized eta invariant of Lott ([**52**],[**68**]):

$$
\langle \tau_{\langle g\rangle}, \mathrm{Ch}^{del}(\rho(\widetilde{D}))\rangle = \eta_{\langle g\rangle}(\widetilde{D})\,. \tag{1.11}
$$

Put it differently,

$$
\eta_{\langle g\rangle}(\widetilde{D}) = \rho_{\tau_{\langle g\rangle}}(\widetilde{D}) \tag{1.12}
$$

which expresses the delocalized eta invariant of Lott as a rho number obtained from the general rho class $\rho(\widetilde{D})$. This result is also discussed in [**18**].

Predecessors of our work are [**33**], where the relative eta invariants $\rho_{\alpha-\beta}(D)$ of Atiyah-Patodi-Singer associated to a virtual representation of Γ of dimension 0 are obtained as pairings of the rho class in $\mathrm{S}^\Gamma_*(\widetilde{M})$ with a trace, i.e. a cyclic cohomology class of degree 0. For recent interesting results on the Atiyah-Patodi-Singer rho invariant see [**3, 4**]. Similarly, [**8**] identifies the Cheeger-Gromov L^2-rho invariant $\rho_{(2)}(\widetilde{D})$ through a pairing of the rho class in $\mathrm{S}^\Gamma_*(\widetilde{M})$ with a trace (the difference of the canonical L^2-trace and the trivial representation), i.e. a cyclic cocycle of degree 0.

These cocycles a priori are defined for a version of the Higson-Roe sequence where one uses a slightly larger completion of $\mathbb{C}\Gamma$ than $C^*_{red}\Gamma$. The maximal completion $C^*_{max}\Gamma$ can of course also be used. For groups where the canonical map

$C^*_{max}\Gamma \to C^*_{red}\Gamma$ induces an isomorphism in K-theory, this fits directly in the context of our paper and is a very special case (with cocycles of degree 0 which are defined on the whole C^*-algebra instead of a smooth subalgebra).

We stress the consequence that thanks to the present work and these contributions of Higson-Roe and Benameur-Roy, the three lower rho invariants introduced earlier in this introduction, that is

$$\rho_{\alpha-\beta}(D), \quad \rho_{(2)}(\widetilde{D}), \quad \eta_{<x>}(\widetilde{D})$$

have all been obtained through the rho class $\rho(\widetilde{D}) \in \mathrm{S}^\Gamma_*(\widetilde{M})$ and its paring with suitable 0-cocycles.

More generally, using our results, the delocalized Atiyah-Patodi-Singer index theorem in K-theory [**65, 66, 90**] and the higher Atiyah-Patodi-Singer index theorem in non-commutative de Rham homology [**44**], we are able to show (if M is a boundary) that

$$\mathrm{Ch}^{del}(\rho(\widetilde{D})) = -\frac{1}{2}\varrho_{\mathrm{Lott}}(\widetilde{D}) \quad \text{in} \quad H^{del}_*(\mathcal{A}\Gamma), \tag{1.13}$$

where on the right hand side the higher rho invariant of Lott appears. This solves a question raised by Lott in [**52**, Remark 4.11, (3)]. Notice that the compatibility with Lott's treatment is also tackled in [**18**, Section 8] where it is established for the corresponding higher rho numbers (under an assumption of polynomial growth of the group, but without the assumption that M is a boundary).

Contributions on higher rho numbers associated to 2-cocycles are given in [**6**]; see also [**5**].

The mapping of the Higson-Roe analytic surgery sequence to a homological sequence was a program that had been already proposed in [**65**]. One might wonder why it took so long to achieve it. The problem is that the algebra $D^*(\widetilde{M})^\Gamma$ is "very large"; what made the difference for the present work is the description given by the third author of the surgery sequence through the adiabatic groupoid [**90**], involving much smaller algebras. Here we use a slight variation of this description, see [**89**]. In [**89**] the compatibility between the Higson-Roe approach and the approach via groupoids is carefully explained.

The mapping of the Higson-Roe surgery sequence to homology already appears in the work of Deeley and Goffeng, see [**23**]. However, their solution passes through the geometric realisation of the Higson-Roe sequence, where geometric means *à la* Baum-Douglas; more precisely, in the work of Deeley-Goffeng one maps out of the Higson-Roe surgery sequence by inverting the isomorphisms going from the geometric realization of the Higson-Roe sequence to the true Higson-Roe surgery sequence. Inverting explicitly these isomorphisms, especially for the part involving $\mathrm{S}^\Gamma_*(\widetilde{M})$, is a difficult problem. Our Chern characters, on the other hand, are computable by direct calculations. It remains an interesting problem to show the full compatibility of the two results.

We finally come to the second paring, the one with the relative cohomology $H^*(M \xrightarrow{u} B\Gamma)$. There is certainly a pairing, obtained via the Chern character

$$H^*(M) \times K_*(M) \to \mathbb{C}\,. \tag{1.14}$$

Let us identify $H^*(\Gamma) = H^*(B\Gamma)$. As already pointed out, if Γ is Gromov hyperbolic or of polynomial growth, we have the Connes-Moscovici pairing, rewritten as

$$H^*(B\Gamma) \times K_*(C^*_{red}\Gamma) \to \mathbb{C}. \tag{1.15}$$

Writing the Higson-Roe analytic surgery sequence with Paschke duality

$$\cdots \to \mathrm{S}^{\Gamma}_{*}(\widetilde{M}) \to K^{\Gamma}_{*}(\widetilde{M}) \xrightarrow{\mu} K_{*}(C^{*}_{red}\Gamma) \to \mathrm{S}^{\Gamma}_{*-1}(\widetilde{M}) \to \cdots$$

and comparing it with the long exact sequence

$$\cdots \leftarrow H^{n+1}(M \xrightarrow{u} B\Gamma) \leftarrow H^{n}(M) \leftarrow H^{n}(B\Gamma) \leftarrow H^{n}(M \xrightarrow{u} B\Gamma) \leftarrow \cdots$$

we understand that $\mathrm{S}^{\Gamma}_{*}(\widetilde{M})$ should be paired with $H^{n}(M \xrightarrow{u} B\Gamma)$ with $* = n$ (mod 2). This is precisely what we achieve in the second part of this article. Crucial for this result is an Alexander-Spanier description of $H^{*}(M \xrightarrow{u} B\Gamma)$, the new description by Zenobi of the group $\mathrm{S}^{\Gamma}_{*}(\widetilde{M})$, together with arguments directly inspired by the work of Connes and Moscovici. We show that the pairing (1.7) exists if Γ satisfies the Rapid Decay (RD) condition and has a combing of polynomial growth (for example, if Γ is Gromov hyperbolic). Similar results through different, more topological, techniques have been discussed in [**79**]. We describe in Section 13 how the two approaches are related.

Finally, we use the pairings for explicit geometric applications. We concentrate on the study of the space $\mathrm{Riem}^{+}(M)$ of metrics of positive scalar curvature. It is well known that for a closed spin manifold M of dimension n which admits a metric g_0 of positive scalar curvature, the higher rho invariant valued in the structure group $\mathrm{S}^{\Gamma}_{n}(\widetilde{M})$ can be used to distinguish components in the space $\mathrm{Riem}^{+}(M)$. The Higson-Roe exact sequence, combined with surgery methods for the construction of new positive scalar curvature metrics, has been used in a quite general context to show that $\pi_0(\mathrm{Riem}^{+}(M))$ is indeed rich, in particular infinite. A somewhat exhaustive culmination of this development is [**84**]. Of course, one can now go on to distinguish higher rho classes in $\mathrm{S}^{\Gamma}_{n}(\widetilde{M})$ also using our pairings with cyclic cocycles.

This is indeed beneficial because it allows to obtain additional information. In particular, we can often control more explicitly how the resulting numeric rho invariants change if one pulls back the metric g by a diffeomorphism $\psi \in \mathrm{Diffeo}(M)$.

This way, we manage to show in numerous cases that the known elements of $\pi_0(\mathrm{Riem}^{+}(M))$ are also in different components of the moduli space, i.e. that $\pi_0(\mathrm{Riem}^{+}(M)/\mathrm{Diffeo}(M))$ is infinite.

As a specific example, we list the following theorem:

THEOREM 1.1. *Let M be a connected closed spin manifold of dimension $n \geq 5$ with fundamental group Γ. Let g_0 be a Riemannian metric of positive scalar curvature on M. Assume that Γ is word hyperbolic or of polynomial growth.*

Assume one of the two conditions: for some $k < n$, $k \equiv n+1 \pmod 4$

- *$e \neq \gamma_0 \in \Gamma$ is an element of finite order, and its centralizer Γ_{γ_0} in Γ satisfies that $H^k(\Gamma_{\gamma_0}, \mathbb{C})$ is one-dimensional*
- *The relative cohomology group $H^k(M \xrightarrow{u} B\Gamma; \mathbb{C})$ is 1-dimensional, where u is a classifying map for the universal covering, i.e. an isomorphism on π_1.*

Then the moduli space of metrics of positive scalar curvature, $\mathrm{Riem}^{+}(M)/\mathrm{Diffeo}(M)$ has infinitely many components.

Notice that we get more general results and more structural information on the moduli space of metrics of positive scalar curvature in Chapters 10, 12, 13 and 14.

The paper is organized as follows. In Chapter 2 and Chapter 3 we recall the alternative description, due to the third author, of the Higson-Roe surgery sequence and the realization of the rho class of an invertible Dirac operator in this new context. This new realisation of the surgery sequence employs Γ-equivariant pseudodifferential operators on $\widetilde{M}$ in an essential way. It also relies heavily on the use of K-theory for relative algebras, i.e. algebra morphisms $f\colon A \to B$. In Chapter 4.1 we recall the definition of Karoubi's Chern character. We develop the theory of non-commutative de Rham homology for relative algebras, and we extend the Chern character to the relative case. In Chapter 5 we discuss results of Lott on pseudodifferential operators on $\widetilde{M}$ with coefficients $\Omega_*(\mathbb{C}\Gamma)$. Chapter 6 is devoted to one of the main results of this paper, the mapping of the analytic surgery sequence to homology. As already remarked, this mapping is valid for any finitely generated discrete group Γ. In this section we also compare the delocalized Chern character of the rho class associated to an invertible Dirac operator $\widetilde{D}$ to Lott's class $\varrho_{\mathrm{Lott}}(\widetilde{D})$ and show that at least if M is a boundary $\mathrm{Ch}^{del}(\rho(\widetilde{D})) = -\frac{1}{2}\varrho_{\mathrm{Lott}}(\widetilde{D}) \in H^{del}_*(\mathcal{A}\Gamma)$. In Chapter 7.6, assuming the group Γ to be Gromov hyperbolic or of polynomial growth, we finally define the higher rho numbers associated to classes in $HC^*_{\mathrm{del}}(\mathbb{C}\Gamma;\langle x\rangle)$. In Chapter 8 we discuss carefully how to realize the relative cohomology groups $H^*(M \xrightarrow{u} B\Gamma)$ in terms of locally zero Γ-equivariant Alexander-Spanier cocycles on $\widetilde{M}$; we use this description and the pseudodifferential realization of $\mathrm{S}^{\Gamma}_*(\widetilde{M})$ in order to define, under suitable assumptions on Γ, the pairing $H^*(M \xrightarrow{u} B\Gamma) \times \mathrm{S}^{\Gamma}_*(\widetilde{M}) \to \mathbb{C}$.

In Chapter 9 we provide explicit formulae for the higher delocalized Atiyah-Patodi-Singer indices of a Galois covering with boundary $\widetilde{W}$, expressing these indices in terms of the higher rho numbers of the boundary $\partial\widetilde{W}$.

In Chapter 11, we give general results on the action of the diffeomorphism group $\mathrm{Diffeo}(M)$ on the whole diagram (1.8) and the resulting naturality of the parings like (1.7) and (1.10). We give explicit geometric applications of this discussion in Chapters 12, 13 and 14.

As this article is rather long, we would like to give at this point a **summary of our main results**:

(1) the development of a relative Chern character in the setting of relative non-commutative de Rham homology, see Chapter 4.2;
(2) the mapping of the analytic surgery exact sequence to non-commutative de Rham homology without any assumption on the fundamental group Γ, see Theorem 6.20;
(3) the isomorphism $HC^*_{pol}(\mathbb{C}\Gamma;\langle x\rangle) \cong HC^*(\mathbb{C}\Gamma;\langle x\rangle)$ for Γ hyperbolic, which says that the cyclic cohomology of the group ring can be calculated using cochains of polynomial growth along each conjugacy class, see Corollary 7.9;
(4) the construction of the pairing $\langle\cdot,\cdot\rangle\colon HC^*(\mathbb{C}\Gamma;\langle x\rangle) \times H^{del}_*(\mathcal{A}\Gamma) \to \mathbb{C}$, $\forall \langle e\rangle \neq \langle x\rangle \in \langle\Gamma\rangle$ for Γ hyperbolic, where $\mathcal{A}\Gamma$ is the Puschnigg algebra;
(5) the construction of the pairing $HC^*(\mathbb{C}\Gamma;\langle x\rangle) \times \mathrm{S}^{\Gamma}_*(\widetilde{M}) \longrightarrow \mathbb{C}$ for Γ hyperbolic, see Theorem 7.22, and the corresponding definition of higher rho numbers given in Definition 7.24;
(6) the description of $H^*(M \xrightarrow{u} B\Gamma)$ using suitable Alexander-Spanier cocycles, via the isomorphism $H^*_{AS,0,\Gamma}(\widetilde{M})^{\Gamma} \xrightarrow{\cong} H^*(M \xrightarrow{u} B\Gamma)$, see Proposition 8.5;

(7) the construction of the pairing $H^*(M \xrightarrow{u} B\Gamma) \times \mathrm{S}^\Gamma_*(\widetilde{M}) \longrightarrow \mathbb{C}$ for discrete groups with property RD and with a combing of polynomial growth, see Theorem 8.24, and the corresponding definition of higher rho numbers given in Definition 8.26;

(8) the treatment of delocalized higher Atiyah-Patodi-Singer index formulae associated to elements in $HC^*(\mathbb{C}\Gamma; \langle x \rangle)$ and $H^*(M \xrightarrow{u} B\Gamma)$;

(9) a detailed discussion about the action of the diffeomorphism group of M on the analytic surgery sequence, the related de Rham non-commutative homology sequence and the compatibility between these two actions;

(10) the discussion of several geometric applications of higher rho numbers, associated either to delocalized cyclic cocycles $HC^*(\mathbb{C}\Gamma; \langle x \rangle)$ or to elements of the relative cohomology group $H^*(M \xrightarrow{u} B\Gamma)$, in problems connected with the classification of metrics of positive scalar curvature, showing in particular many new cases in which the moduli space of such metrics has infinitely many components. More precisely, we provide strong lower bounds on the rank of the moduli space in terms of delocalized and relative group cohomology. This rank is defined via the group of coinvariants for the usual group structure on the space of concordance classes of metrics of positive scalar curvature.

The results of this paper relative to hyperbolic groups were announced in a seminar at Fudan University in January 2019. At the same time the article by Chen, Wang, Xie and Yu [**18**] appeared, giving, among other things, a different treatment of some of the same results, most notably the well-definedness of the higher rho numbers associated to elements in $HC^*(\mathbb{C}\Gamma, \langle x \rangle)$. The two articles have a non-trivial intersection but also a non-trivial symmetric difference. For example, in [**18**], building on [**83**], the convergence of Lott's delocalized eta invariant is established under very general assumptions; we don't discuss at all this problem in this paper. On the other hand, items (2), (6), (7) and (10) in the summary given above are results treated here and not in [**18**]. Moreover, we believe that our results in (3) and (4) are more precise than those in [**18**, Section 5] and derived in a systematic and complete way using homological algebra methods throughout.

Notice that the techniques employed for the common results are completely different. Recently the results of [**18**] have been extended to the case of discrete groups of polynomial growth, see [**37**]. In this second version of our paper we observed that our arguments work equally well for the (easier) case of groups of polynomial growth.

Acknowledgments

We are indebted to Nigel Higson for suggesting the pairing (1.7) and Proposition 8.5 (which we therefore call "Higson's Lemma"). We thank Michael Puschnigg for very useful discussions about hyperbolic groups. We also had interesting correspondence and discussions with John Lott, Ralf Meyer and Ryszard Nest and we take this opportunity to thank them all. Finally, we thank the referee of this article for useful comments and suggestions.

Part of this work was performed while the second author was visiting Sapienza Università di Roma, with a 3-months visiting professorship. There were also several short-time visits of Schick and Zenobi to Sapienza. We are grateful to Sapienza Università di Roma for the financial support that made all these visits possible. The first and third author have been supported by the DFG-SPP 2026 "Geometry at Infinity".

CHAPTER 2

The relation between the structure algebra $D^*(\widetilde{M})^\Gamma$ and $\Psi^0_\Gamma(\widetilde{M})$

Let $\Psi^0_{\Gamma,c}(\widetilde{M})$ be the algebra of Γ-equivariant pseudodifferential operators of order 0 on $\widetilde{M}$ with Schwartz kernel of Γ-compact support. Let $\Psi^0_\Gamma(\widetilde{M})$ denote the C^*-closure of $\Psi^0_{\Gamma,c}(\widetilde{M})$ in the bounded operators of $L^2(\widetilde{M})$. Recall the algebra $D^*_c(\widetilde{M})^\Gamma$ of finite propagation pseudolocal bounded operators on $L^2(\widetilde{M})$ and its C^*-closure $D^*(\widetilde{M})^\Gamma$. We don't repeat the details of the definitions of these algebras, as we are actually not using them. Directly from the definition of $D^*_c(\widetilde{M})^\Gamma$ and the properties of pseudodifferential operators, we understand that if $P \in \Psi^0_{\Gamma,c}(\widetilde{M})$ then $P \in D^*_c(\widetilde{M})^\Gamma$. One might then wonder what is the precise relationship between the algebra $\Psi^0_\Gamma(\widetilde{M})$ and the algebra $D^*(\widetilde{M})^\Gamma$, at least as far as K-theory is concerned. The next subsection, based on [**89**], gives an answer to this question.

We shall consider the mapping cone C*-algebra of a *-homomorphism $\varphi\colon A \to B$. We will use both notations $K_*(A \xrightarrow{\varphi} B)$ and $K_*(\varphi)$ for denoting the K-theory of this mapping cone. This is also called the relative K-theory of φ. In the realization of $K_*(\varphi)$ as in [**3**, Section 2.3], elements of $K_0(\varphi)$ are given by homotopy classes of triples $[p_0, p_1; q_t]$ with p_0, p_1 projections over A and q_t a path of projections over B such that $\varphi(p_i) = q_i$ for $i = 0, 1$. Furthermore, elements in $K_1(\varphi)$ are couples $[u, f]$ with u a unitary over A and f a path of unitaries over B joining $\varphi(u)$ and the identity.

2.1. Realizing the surgery sequence with pseudodifferential operators

Consider the classic short exact sequence of pseudodifferential operators

$$0 \to C^*_{red}(\widetilde{M} \times_\Gamma \widetilde{M}) \to \Psi^0_\Gamma(\widetilde{M}) \xrightarrow{\sigma} C(S^*M) \to 0$$

where we recall that $\Psi^0_\Gamma(\widetilde{M})$ denotes the C^*-closure of $\Psi^0_{\Gamma,c}(\widetilde{M})$ in the bounded operators of $L^2(\widetilde{M})$, σ is the principal symbol map and $C^*_{red}(\widetilde{M} \times_\Gamma \widetilde{M})$ is the C*-closure of $\Psi^{-\infty}_{\Gamma,c}(\widetilde{M})$, the ideal of smoothing Γ-equivariant operators with Γ-compact support. Consider the natural inclusion $C(M) \xrightarrow{\mathfrak{m}} \Psi^0_\Gamma(\widetilde{M})$, with $\mathfrak{m}$ denoting the map that associates to $f \in C(M)$ the multiplication operator by the equivariant lift $\tilde{f}$ of f. This induces a long exact sequence in K-theory for the mapping cone C^*-algebras:

$$\text{(2.1)}\quad \cdots \xrightarrow{\partial} K_*(0 \hookrightarrow C^*_{red}(\widetilde{M} \times_\Gamma \widetilde{M})) \xrightarrow{i_*} K_*(C(M) \xrightarrow{\mathfrak{m}} \Psi^0_\Gamma(\widetilde{M})) \xrightarrow{\sigma_*} K_*(C(M) \xrightarrow{\pi^*} C(S^*M)) \xrightarrow{\partial} \cdots$$

where $\pi\colon S^*M \to M$ is the bundle projection of the cosphere bundle. Observe that $K_*(0 \hookrightarrow C^*_{red}(\widetilde{M}\times_\Gamma \widetilde{M}))$ is nothing but $K_*(C^*_{red}(\widetilde{M}\times_\Gamma \widetilde{M})\otimes C_0(0,1))$.

In [**89**] the author proves the existence of a canonical isomorphism of long exact sequences

(2.2)

$$\begin{array}{ccccc} \to K_{*+1}(C^*(\widetilde{M})^\Gamma) & \longrightarrow & K_{*+1}(D^*(\widetilde{M})^\Gamma) & \longrightarrow & K_{*+1}(D^*(\widetilde{M})^\Gamma/C^*(\widetilde{M})^\Gamma) \to \\ \downarrow\cong & & \downarrow\cong & & \downarrow\cong \\ \to K_*(C^*_{red}(\widetilde{M}\times_\Gamma \widetilde{M})\otimes C_0(0,1)) & \longrightarrow & K_*(C(M)\xrightarrow{\mathfrak{m}}\Psi^0_\Gamma(\widetilde{M})) & \longrightarrow & K_*(C(M)\xrightarrow{\pi^*}C(S^*M)) \to \end{array}$$

We remark that the mapping cone C^*-algebra $\mathcal{C}_{\pi^*}$, that is $C(M)\xrightarrow{\pi^*}C(S^*M)$, is isomorphic to $C_0(T^*M)$. Indeed recall that, if $f\colon X\to Y$ is a continuous map of topological spaces, then the mapping cone of f is the quotient of $X\times[0,1]\sqcup Y$ by the relation $(x,1)\sim f(x)$ and $(x,0)\sim(x',0)$. It is easy to see that the algebra of continuous functions on the mapping cone of f vanishing at $[X\times\{0\}]$ is isomorphic to the mapping cone C*-algebra of $f^*\colon C(Y)\to C(X)$. Since the mapping cone of π is the one-point compactification of T^*X, one has that the mapping cone C*-algebra associated to π^* is isomorphic to $C_0(T^*M)$ as desired.

Recall that to every C^*-algebra homomorphism we obtain an associated long exact K-theory sequence using the mapping cone, which we call the pair sequence. For $\mathfrak{m}\colon C(M)\to\Psi^0_\Gamma(\widetilde{M})$ this gives, using the abbreviation $K_*(\mathfrak{m})$ for the K-theory of the mapping cone introduced above

$$\cdots\to K_*(C(M)\otimes C_0(0,1))\xrightarrow{\mathfrak{m}\otimes\mathrm{id}_*}K_*(\Psi^0_\Gamma(\widetilde{M})\otimes C_0(0,1))$$
$$\to K_*(\mathfrak{m})\xrightarrow{(ev_0)_*}K_*(C(M))\to\cdots$$

where the map $(ev_0)_*$ is given by the following composition

$$K_*(\mathfrak{m})\longrightarrow K_*(\pi^*)\xrightarrow[\cong]{}K_*(C_0(T^*M))\xrightarrow{j_*}K_*(C(D^*M))\xrightarrow[\cong]{i_*}K_*(C(M))\,.$$

Here, $\pi^*\colon C(M)\to C(S^*M)$ is the pull-back map through the bundle projection, $j\colon T^*M\to D^*M$ is given by a diffeomorphism between the cotangent bundle and the open codisk bundle, and $i\colon M\to D^*M$ is the inclusion. If the Euler characteristic of M is zero, then j^* is the 0 map and, consequently, so is $(ev_0)_*$.

Let us prove that $j_*=0$ if $\chi(M)=0$: we consider the following exact sequence

(2.3)

$$\cdots\longrightarrow K_*(C_0(T^*M))\xrightarrow{j_*}K_*(C(D^*M))\longrightarrow K_*(C(S^*M))\longrightarrow\cdots.$$

Then it is easy to see that its boundary map is given by the composition of the suspension isomorphism $K_*(C(S^*M))\to K_{*+1}(C_0(T^*M\setminus M))$ and the inclusion $\iota\colon C_0(T^*M\setminus M)\to C_0(T^*M)$. Indeed, if the Euler characteristic of M is zero, the cotangent bundle of M admits a nowhere vanishing section ξ. Then take a diffeomorphism φ between T^*M and the open disc bundle $\mathring{D}^*M$ and a diffeomorphism ψ between $\mathring{D}^*M$ and U_ξ, a tubular neighbourhood of the image of ξ. The morphism $h=\psi\circ\varphi$ induces a morphism $h_*\colon C_0(T^*M)\to C_0(T^*M\setminus M)$ such that $\iota_*\circ h_*=\mathrm{Id}\colon K_*(C_0(T^*M))\to K_*(C_0(T^*M))$. This means that ι_* is surjective and, by exactness of (2.3), that j_* is zero.

Summarizing: if the Euler characteristic is zero then $j_* = 0$ and consequently $(ev_0)_* = 0$.

This tells us that $K_*(\Psi^0_\Gamma(\widetilde{M})) \to K_{*+1}(\mathfrak{m})$ is surjective and then any class in $K_{*+1}(\mathfrak{m})$ can be lifted, up to Bott periodicity, to $K_*(\Psi^0_\Gamma(\widetilde{M}))$. Observe that this is always true for $*+1 = \dim M$ odd. Thus, when the Euler characteristic of M is zero, we have the following surjective morphism of exact sequences:

(2.4)

$$\begin{array}{ccccc} \to K_{*+1}(C^*_{red}(\widetilde{M}\times_\Gamma \widetilde{M})) & \longrightarrow & K_{*+1}(\Psi^0_\Gamma(\widetilde{M})) & \longrightarrow & K_{*+1}(C(S^*M)) \to \\ \downarrow & & \downarrow & & \downarrow \\ \to K_*(C^*_{red}(\widetilde{M}\times_\Gamma \widetilde{M})\otimes C_0(0,1)) & \longrightarrow & K_*(C(M) \xrightarrow{\mathfrak{m}} \Psi^0_\Gamma(\widetilde{M})) & \longrightarrow & K_*(C(M) \xrightarrow{\pi^*} C(S^*M)) \to \end{array}$$

We shall use this information in order to simplify some of our formulae.

CHAPTER 3

Rho classes

3.1. Analytic properties of $\pi_{\geq}(\widetilde{D})$

Assume that $\widetilde{D}$ is a Γ-invariant Dirac type operator on $\widetilde{M}$ with coefficient in a Clifford bundle E, which will be omitted from the notation all along this section. Assume that $\widetilde{D}$ is L^2-invertible. Let $\pi_{\geq}\colon \mathbb{R} \to \mathbb{R}$ be the characteristic function of $[0, \infty)$. Functional calculus then allows us to define $\pi_{\geq}(\widetilde{D})$ as a bounded operator on $L^2(\widetilde{M})$. We shall need more precise properties of this operator and these are given in the next two propositions.

PROPOSITION 3.1. *The bounded operator $\pi_{\geq}(\widetilde{D})$ is an element in the C^*-closure $\Psi^0_\Gamma(\widetilde{M})$ of $\Psi^0_{\Gamma,c}(\widetilde{M})$.*

PROOF. First recall that there is a chopping function $f\colon \mathbb{R} \to [-1, 1]$ which is odd, monotonically increasing, smooth, "quasi-Schwarz", with $\lim_{t\to\infty} f(t) = 1$ and such that its Fourier transform has compact support. Note that this Fourier transform has a singularity at zero and is smooth outside. Then $f(\widetilde{D})$ is a pseudo-differential operator of order zero, which is Γ-compactly supported because of finite propagation speed and the compact support of $\hat{f}$. By the mapping properties of f, $\pi_{\geq}(\widetilde{D}) = \pi_{\geq}(f(\widetilde{D}))$. Because C^*-algebras are closed under continuous functional calculus and because $f(\widetilde{D})$ is invertible, $\pi_{\geq}(f(\widetilde{D}))$ is contained in the C^*-closure of $\Psi^0_{\Gamma,c}(\widetilde{M})$. □

PROPOSITION 3.2. *If $\widetilde{D}$ is L^2-invertible, then $\pi_{\geq}(\widetilde{D})$ belongs to any holomorphically closed *-algebra $\mathcal{A}$ such that*

$$\Psi^0_{\Gamma,c}(\widetilde{M}) \subset \mathcal{A} \subset \Psi^0_\Gamma(\widetilde{M}).$$

PROOF. With the choice of f as in the previous proof, so that $f(\widetilde{D}) \in \Psi^0_{\Gamma,c}(\widetilde{M})$, we observe that $\pi_{\geq}(f(\widetilde{D}))$ is obtained by holomorphic functional calculus; indeed for any $T \in \Psi^0_{\Gamma,c}(\widetilde{M})$ invertible in $\Psi^0_\Gamma(\widetilde{M})$ we have

$$\pi_{\geq}(T) = \frac{(T \circ (\sqrt{T^*T})^{-1}) + 1}{2}.$$

It then follows that $\pi_{\geq}(f(\widetilde{D}))$ belongs to any algebra closed under holomorphic functional calculus which contains $\Psi^0_{\Gamma,c}(\widetilde{M})$. □

REMARK 3.3. The same reasoning applies to the operator defined by the sign function, $\mathrm{sign}(\widetilde{D}) = 2\pi_{\geq}(\widetilde{D}) - 1$.

3.2. Rho classes and K-theory of pseudodifferential operators

We want to give the new realisation of the rho classes in the setting of relative K-theory, see [**89**]. Concretely, cycles for $K_0(A \xrightarrow{f} B)$ are triples (p_0, p_1, q_t) where p_0, p_1 are idempotent matrices over A and $(q_t)_{t\in[0,1]}$ is a continuous loop of idempotent matrices in B from $f(p_0)$ to $f(p_1)$. Cycles for $K_1(A \xrightarrow{f} B)$ are of the form (u, v_t) where u is an invertible matrix over A and $(v_z)_{z\in S^1\subset\mathbb{C}}$ is a loop of invertible matrices in B with $v_1 = f(u)$. Let $\widetilde{M}$ be a Galois Γ-covering of a closed smooth manifold M. Let $\widetilde{D}$ be an L^2-invertible Γ-equivariant Dirac operator on $\widetilde{M}$.

DEFINITION 3.4.

- If $\dim M$ is even, then $\varrho(\widetilde{D})$ is defined as the class in $K_0(\mathfrak{m})$ given by

(3.1)
$$\left[\begin{pmatrix}1_+ & 0\\ 0 & 0\end{pmatrix}, \begin{pmatrix}0 & 0\\ 0 & 1_-\end{pmatrix}; \begin{pmatrix}\cos^2\left(\frac{\pi}{2}t\right)1_+ & \cos\left(\frac{\pi}{2}t\right)\sin\left(\frac{\pi}{2}t\right)\operatorname{sgn}(\widetilde{D})_-\\ \cos\left(\frac{\pi}{2}t\right)\sin\left(\frac{\pi}{2}t\right)\operatorname{sgn}(\widetilde{D})_+ & \sin^2\left(\frac{\pi}{2}t\right)1_-\end{pmatrix}\right]$$

  where 1_+ is the projection which defines the $C(M)$-module $C(M, E_+)$ of continuous sections of the even part E_+ of the Clifford bundle E, entering into the definition of $\widetilde{D}$, and 1_- is defined correspondingly using the odd part of E.
- If $\dim \widetilde{M}$ is odd then $\varrho(\widetilde{D})$ is defined as the class in $K_1(\mathfrak{m})$ given by $[1, 1 + \pi_{\geq}(\widetilde{D})(z-1)]$, with $z \in S^1 \subset \mathbb{C}$.

REMARK 3.5. It is proved in [**89**] that these definitions are compatible with the original ones, as given in [**65**], through the isomorphism (2.2). Notice that in the odd dimensional case we know that $K_1(\Psi^0_\Gamma(\widetilde{M}) \otimes C_0(0,1)) \to K_1(\mathfrak{m})$ is surjective. In fact, the rho class we have just defined is the image, up to suspension isomorphism, of the class in $K_0(\Psi^0_\Gamma(\widetilde{M}))$ given by $[\pi_\geq]$. Put it differently, $[\pi_\geq] \in K_0(\Psi^0_\Gamma(\widetilde{M}))$ is a natural lift of $\varrho(\widetilde{D}) \in K_1(\mathfrak{m})$. In the sequel, we shall be interested in pairing $\varrho(\widetilde{D}) \in K_1(\mathfrak{m})$ with suitable delocalized cyclic cocycles. Since the kernel of $K_1(\Psi^0_\Gamma(\widetilde{M}) \otimes C_0(0,1)) \to K_1(\mathfrak{m})$ is $K_0(C(M))$ (up to suspension isomorphism) and since this group pairs trivially with such delocalized cocycles, we conclude that in the odd dimensional case we can focus solely on $K_0(\Psi^0_\Gamma(\widetilde{M}))$ and on $[\pi_\geq] \in K_0(\Psi^0_\Gamma(\widetilde{M}))$.

DEFINITION 3.6. Let g be a Γ-equivariant metric with positive scalar curvature on a spin Γ-covering $\widetilde{M}$ of M. Then we will denote by $\varrho(g) \in K_*(\mathfrak{m})$ the rho class associated to the spin Dirac operator $\widetilde{D}$ defined on the spinor bundle. Recall that it is L^2-invertible because the scalar curvature is bounded below by a positive constant.

CHAPTER 4

Non-commutative de Rham homology and Chern characters

4.1. Non-commutative de Rham homology

For the material in this section we refer the reader to [**40**]. We recall the standard definitions which lead to the de Rham cohomology of a Fréchet algebra, the target of the Chern character from the K-theory of the algebra.

Let A be a Fréchet algebra, not necessarily unital. Then we denote by A_+ its unitalization, i.e. $A_+ := A \oplus \mathbb{C}$ with multiplication such that $1 \in \mathbb{C}$ is the unit of A_+. Note that, if A used to have a unit $e \in A$, $e \in A_+$ is a central idempotent and we have the algebra decomposition $A_+ = eA \oplus (1-e)\mathbb{C}$. We always have the split inclusion $\mathbb{C} \hookrightarrow A_+$ of the projection $A_+ = A \oplus \mathbb{C} \to \mathbb{C}$ which is an algebra homomorphism.

A Fréchet differential graded algebra, shortly a DGA, $(\Omega_\bullet, d)$, with $\Omega_\bullet = \bigoplus \Omega_i$, is said to be a DGA over A if $\Omega_0 = A$. Let us denote by $\Omega_\bullet^+$ the unitalization of $\Omega_\bullet$, it is a DGA over A_+ given by $\Omega_\bullet^+ := \Omega_\bullet \oplus \Omega_\bullet(\mathbb{C})$. Because $\Omega_\bullet(\mathbb{C})$ is isomorphic to $\mathbb{C}$ concentrated in degree 0, the inclusion $\mathbb{C} \to A_+$ always induces a split injection of DGAs $\Omega_*(\mathbb{C}) \hookrightarrow \Omega_\bullet^+$.

EXAMPLE 4.1. Given a unital algebra B over $\mathbb{C}$ with unit e, the universal differential graded algebra $\Omega_\bullet(B)$ is defined as $\Omega_n(B) = B \otimes \bar{B}^{\otimes n}$ with $\bar{B} = B/\mathbb{C}e$ the quotient by the multiples of the identity. The differential sends $b \in B$ to $db := 1 \otimes [b] \in \bar{B}$. The product is defined by concatenation and the Leibniz rule. Elements are linear combinations of the form $b_0 db_1 \dots db_n$.

If A is a unital Fréchet algebra, we denote by $\widehat{\Omega}_\bullet(A)$ the corresponding degree-wise completed object, constructed using the projective tensor product.

DEFINITION 4.2. Let us consider a DGA $\Omega_\bullet$ over a (Fréchet) algebra A, not necessarily unital. Let $\Omega_{ab,\bullet}$ denote the vector space quotient of $\Omega_\bullet$ by the closure of the subspace generated by the graded commutators $\overline{[\Omega_\bullet, \Omega_\bullet]}$. Notice that d induces a differential on $\Omega_{ab,\bullet}$ since $d[\omega_i, \omega_j] = [d\omega_i, \omega_j] + (-1)^i[\omega_i, d\omega_j]$ for $\omega_i \in \Omega_i$, $\omega_j \in \Omega_j$. Define the de Rham homology $H_*(\Omega_\bullet, d)$ as the i-th homology group of the complex $(\Omega_{ab,\bullet}, d)$.

REMARK 4.3. To be able to apply standard homological algebra, we use algebraic homology, i.e. we quotient by the image of the differential, and not its closure. Here, our conventions differ from part of the literature. However, there is the canonical map from algebraic homology to the version where one divides by closures, and our invariants are lifts of the previously constructed ones (where applicable), and thus contain a priori even more information. By mapping to the quotient by closures we obtain the usual invariants.

DEFINITION 4.4. We define the *universal de Rham homology* of A

$$H_*(A) := \operatorname{coker}\left(H_*(\widehat{\Omega}_\bullet(\mathbb{C})) \to H_*(\widehat{\Omega}_\bullet(A_+))\right).$$

Moreover, for a DGA $\Omega_\bullet$ over a Fréchet algebra A, we define the *augmented homology*

$$H_*^+(\Omega_\bullet) := \operatorname{coker}(H_*(\mathbb{C}) \to H_*(\Omega_\bullet^+)),$$

which is canonically a summand of $H_*(\Omega_\bullet^+)$.

REMARK 4.5. Notice that, if $\Omega_\bullet$ is a DGA over A, the canonical map $H_*^+(\Omega_\bullet) \to H_*(\Omega_\bullet)$ is an isomorphism. Because of this fact we will freely use the non augmented version also when, to make sense of a situation, the augmentation seems necessary. Furthermore, observe that, by universality, we always have a map of complexes $\theta\colon \widehat{\Omega}_\bullet(A_+) \to \Omega_\bullet^+$ for any Fréchet DGA $\Omega_\bullet^+$ over A_+ compatible with the map from $\Omega_\bullet(\mathbb{C})$. Consequently, one always has a map

$$\theta_*\colon H_*(A) \to H_*^+(\Omega_\bullet) \cong H_*(\Omega_\bullet).$$

Now consider a homomorphism of Fréchet algebras $\varphi\colon A \to B$. We will work with *relative* non-commutative de Rham homology in this context and work out the basics for this now. We denote $\varphi_+\colon A_+ \to B_+$ the canonical unital homomorphism of the unitalizations.

DEFINITION 4.6. A *pair of DGAs over* φ is given by a homomorphism of DGAs $\Phi_\bullet\colon \Omega_\bullet^A \to \Omega_\bullet^B$ with (Ω^A, d^A) a DGA over A, (Ω^B, d^B) a DGA over B and $\Phi_0 = \varphi$. The *universal pair over* φ is given by the homomorphism of DGAs $\Omega_\bullet(\varphi_+)\colon \widehat{\Omega}_\bullet(A_+) \to \widehat{\Omega}_\bullet(B_+)$, obtained by the universal property of $\widehat{\Omega}_\bullet(A_+)$.

By general principles, we can associate to a pair over φ the mapping cone complex

$$(\Omega^A \xrightarrow{\Phi} \Omega^B)_{ab,\bullet} := \Omega^A_{ab,\bullet+1} \oplus \Omega^B_{ab,\bullet}$$

which is equipped with the differential

$$d^\Phi := \begin{pmatrix} d^A & 0 \\ -\Phi_{\bullet+1} & d^B \end{pmatrix}.$$

DEFINITION 4.7. Define the *relative homology* of the pair $\Phi_\bullet\colon \Omega_\bullet^A \to \Omega_\bullet^B$ as the homology of the mapping cone complex and denote it by $H_*(\Phi)$. The *universal relative de Rham homology* of φ is the homology of the mapping cone complex of $\Omega_\bullet(\varphi_+)$ and is denoted by $H_*(\varphi)$.

It is a standard fact in homological algebra that a pair $\Phi_\bullet\colon \Omega_\bullet^A \to \Omega_\bullet^B$ over $\varphi\colon A \to B$ gives rise to a long exact sequence of homology groups

(4.1)

$$\cdots \longrightarrow H_*(\Omega_\bullet^B) \longrightarrow H_*(\Phi) \longrightarrow H_{*+1}(\Omega_\bullet^A) \xrightarrow{\Phi_*} H_{*+1}(\Omega_\bullet^B) \longrightarrow \cdots$$

Moreover, by naturality, we can associate to φ a long exact sequence of universal de Rham homology groups

$$\cdots \longrightarrow H_*(B) \longrightarrow H_*(\varphi) \longrightarrow H_{*+1}(A) \xrightarrow{\varphi_*} H_{*+1}(B) \longrightarrow \cdots \tag{4.2}$$

Observe that if $\Phi_\bullet^+\colon \Omega_\bullet^{A,+} \to \Omega_\bullet^{B,+}$ is the augmentation of $\Phi_\bullet$, then $H_*(\Phi^+) \cong H_*(\Phi)$. Thanks to this fact and to Remark 4.5, there always exists a canonical natural mapping from (4.2) to (4.1).

REMARK 4.8. In contrast to some of the literature, for what concern the universal de Rham homology, we use here the unitalization and augmented versions. This has the advantage to be applicable to non-unital algebras. Moreover, it is perfectly in line with the definition of K-theory. Finally, we argue that $H_*^+(A)$ is more appropriate also because it canonically maps as a subgroup into cyclic homology, the other popular homology theory for algebras, but it only works in the augmented setting.

4.2. Chern characters

Let $(\Omega_\bullet, d)$ be a DGA over A. We are going to recall the definition of the Chern character

$$\mathrm{Ch}\colon K_0(A) \to H_{\mathrm{even}}(\Omega_\bullet). \tag{4.3}$$

All along this section we do not want to care about the fact that A is unital or not. Even if the general definition $K_0(A) = \mathrm{coker}(K_0(\mathbb{C}) \to K_0(A_+))$ for K-theory groups would consequently imply that we use the augmented version of the homology groups, we will instead employ the non augmented notation, bearing in mind that Remark 4.5 prevents us from any risk of inconsistency.

For this, consider a class $[p] \in K_0(A)$, where p is an idempotent $n \times n$ matrix over A, and denote by E the finitely generated projective module pA^n. Then $\nabla := p \circ d \circ p\colon E \to E \otimes_A \Omega_1$ is a connection and its curvature $\Theta := \nabla^2$ is an A-linear endomorphism of E with coefficients in Ω_2 explicitly given by $pdpdp$.

DEFINITION 4.9. The degree $2k$ part of the Chern character of $[p] \in K_0(A)$ associated to the DGA $(\Omega_\bullet, d)$ is defined as the class

$$\mathrm{Ch}_{2k}([p]) := \frac{1}{k!}\,\mathrm{Tr}(\Theta^k) \in H_{2k}(\Omega_\bullet).$$

We want to define a relative version of the Chern character. To this end we need to prove a transgression formula for the absolute Chern character. Let p_t be a smooth path of idempotents in $M_n(A)$, with $t \in [0,1]$. We regard it as an idempotent $\tilde{p}$ over $C^\infty[0,1]\widehat{\otimes} A$. Moreover, if (Ω_*, d) is a Fréchet DGA over A, then we can canonically associate to it the following Fréchet DGA over $C^\infty[0,1]\widehat{\otimes} A$

$$\left((C^\infty[0,1] \oplus C^\infty[0,1]dt)\widehat{\otimes}\Omega_*,\; D := dt \wedge \frac{\partial}{\partial t} \otimes \mathrm{id} + \mathrm{id} \otimes d\right).$$

The connection $\tilde{p} \circ D \circ \tilde{p}$ is equal to

$$p_t \circ d \circ p_t + p_t \circ dt \wedge \frac{\partial}{\partial t} \circ p_t \tag{4.4}$$

and its curvature $(\tilde{p} \circ D \circ \tilde{p})^2$ is equal to

$$\Theta_t + \Xi_t \wedge dt$$

for some path of 1-forms Ξ_t and with Θ_t being the curvature of the connection $p_t \circ d \circ p_t$.

LEMMA 4.10. *The path Ξ_t of A-linear morphisms with coefficient in Ω_1 is given by $p_t dp_t \dot{p}_t + p_t \dot{p}_t dp_t$.*

PROOF. First recall that the equality $p_t = p_t^2$ implies that $\dot{p}_t(1-p_t) = p_t\dot{p}_t$. It follows that $p_t\dot{p}_tp_t = 0$ and that $0 = d(p_t\dot{p}_tp_t) = d(p_t\dot{p}_t)p_t + p_t\dot{p}_tdp_t$. By a direct calculation we see that

$$(4.5)\quad \begin{aligned}(\tilde{p}\circ D\circ\tilde{p})^2 - \Theta_t &= p_t\circ d\circ p_t\circ dt\wedge\frac{\partial}{\partial t}\circ p_t + p_t\circ dt\wedge\frac{\partial}{\partial t}\circ p_t\circ d\circ p_t\\ &= (p_t\circ d\circ p_t\circ\frac{\partial}{\partial t}\circ p_t + p_t\circ\frac{\partial}{\partial t}\circ p_t\circ d\circ p_t)\wedge dt.\end{aligned}$$

Now apply the coefficient of dt in the last line to an element $x_t \in \tilde{p}(C^\infty[0,1]\otimes A_+^n)$. Then

$$(4.6)\quad \begin{aligned}&p_td(p_t\frac{\partial}{\partial t}(p_tx_t)) + p_t\frac{\partial}{\partial t}(p_td(p_tx_t))\\ &= p_td(p_t(\dot{p}_tx_t + p_t\dot{x}_t)) + p_t\frac{\partial}{\partial t}(p_tdp_tx_t + p_tdx_t)\\ &= p_tdp_t(\dot{p}_tx_t + p_t\dot{x}_t) + p_td(\dot{p}_tx_t + p_t\dot{x}_t)\\ &\quad + p_t(\dot{p}_tdp_tx_t + p_td\dot{p}_tx_t + p_tdp_t\dot{x}_t + \dot{p}_tdx_t + p_td\dot{x}_t)\\ &= (p_tdp_t\dot{p}_t + p_t\dot{p}_tdp_t)x_t\end{aligned}$$

and the result follows. □

It follows that the $2k$-degree component of the Chern character of $\tilde{p}$ is of the following form

$$(4.7)\quad \mathrm{Ch}_{2k}(\tilde{p}) = \mathrm{Ch}_{2k}(p_t) + \overline{\mathrm{Ch}}_{2k-1}(p_t)\wedge dt.$$

We call $\overline{\mathrm{Ch}}_{2k-1}(p_t)$ in the above formula the *transgression Chern character.*

LEMMA 4.11. *The transgression Chern character is explicitly given by*

$$(4.8)\quad \overline{\mathrm{Ch}}_{2k-1}(p_t) = \frac{1}{k!}\,\mathrm{Tr}\left(\Theta_t^{k-1}\Xi_t\right)$$

and satisfies the transgression formula

$$(4.9)\quad \frac{\partial}{\partial t}\mathrm{Ch}_{2k}(p_t) = -d\overline{\mathrm{Ch}}_{2k-1}(p_t).$$

It is functorial for algebra maps $f\colon A\to B$ covered by a DGA map $F\colon \Omega_\bullet^A\to\Omega_\bullet^B$ over $f\colon A\to B$.

PROOF. The first statement is given by a direct calculation of $\mathrm{Ch}_{2k}(\tilde{p})$ as follows

$$\begin{aligned}&\frac{1}{k!}\,\mathrm{Tr}\left((\Theta_t + \Xi_t\wedge dt)^k\right)\\ &= \frac{1}{k!}\,\mathrm{Tr}\left(\Theta_t^k + \sum_{j=0}^{k-1}\Theta_t^j(\Xi_t\wedge dt)\Theta^{k-j-1}\right)\\ &= \frac{1}{k!}\,\mathrm{Tr}(\Theta_t^k) + \frac{1}{(k-1)!}\,\mathrm{Tr}(\Xi_t\Theta_t^{k-1})\wedge dt.\end{aligned}$$

Here we used the fact that Tr is linear and zero on graded commutators. Moreover, since $\mathrm{Ch}_{2k}(\tilde{p})$ is closed with respect to the differential D and $\mathrm{Ch}_{2k}(p_t)$ is closed with respect to d, applying D to (4.7) we obtain (4.9). Functoriality follows directly from (4.8). □

COROLLARY 4.12. *For the smooth path of idempotents p_t we have*

$$\mathrm{Ch}_{2k}(p_0) - \mathrm{Ch}_{2k}(p_1) = d\int_0^1 \overline{\mathrm{Ch}}_{2k-1}(p_t)dt. \tag{4.10}$$

Now we are ready to define the relative Chern character. Recall that the relative K-group $K_0(\varphi)$ is defined via homotopy classes of triples $(p_0, p_1; q_t)$ where p_0 and p_1 are idempotents over A_+ and q_t is a path of idempotents over B_+ such that $q_0 = \varphi(p_0)$ and $q_1 = \varphi(p_1)$. The equality (4.10) allows to give the following definition.

DEFINITION 4.13. The degree $2k-1$ part of the relative Chern character of $[p_0, p_1; q_t] \in K_0(\varphi)$ associated to a pair of DGAs $\Phi_\bullet\colon \Omega^A_\bullet \to \Omega^B_\bullet$ over φ is defined as

(4.11)

$$\mathrm{Ch}^{rel}_{2k-1}([p_0, p_1; q_t]) := \left(\mathrm{Ch}_{2k}(p_0) - \mathrm{Ch}_{2k}(p_1), \int_0^1 \overline{\mathrm{Ch}}_{2k-1}(q_t)\, dt\right) \in H_{2k-1}(\Phi).$$

PROPOSITION 4.14. *The relative Chern character of a class in K-theory does not depend on the choice of a representative and it is therefore well defined.*

PROOF. Let $(p_0^s, p_1^s; q_t^s)$ be a smooth path of relative K-cycles for φ. We can see q_t^s as an idempotent $\tilde{q}$ over $C^\infty[0,1]^2 \otimes B_+$. Consider the Fréchet DGA $\big(C^\infty[0,1]^2 \otimes (1 \oplus dt \oplus ds \oplus dt \wedge ds)\big) \otimes \Omega^B_\bullet$ with differential

$$D := ds \wedge \frac{\partial}{\partial s} + dt \wedge \frac{\partial}{\partial t} + d^B.$$

Let E be the module associated to $\tilde{q}$ and let us equip E with the connection $\tilde{q}D\tilde{q}$. Its curvature is given by

$$\tilde{\Theta} = (q_t^s ds \wedge \frac{\partial}{\partial s} q_t^s + q_t^s dt \wedge \frac{\partial}{\partial t} q_t^s + q_t^s d^B q_t^s)^2$$

which is the sum of four terms

$$\overline{\Theta} = \Theta_{t,s} + \Xi_{t,s} \wedge dt + \Sigma_{t,s} \wedge ds + \Psi_{t,s} \wedge dt \wedge ds.$$

Then the degree $2k$ component of the Chern character of $\tilde{q}$ is given by

$$\mathrm{Ch}_{2k}(\tilde{q}) = \mathrm{Ch}_{2k}(q_t^s) + \overline{\mathrm{Ch}}^t_{2k-1}(q_t^s) \wedge dt + \overline{\mathrm{Ch}}^s_{2k-1}(q_t^s) \wedge ds + \overline{\overline{\mathrm{Ch}}}_{2k-2}(q_t^s) \wedge dt \wedge ds \tag{4.12}$$

where $\overline{\mathrm{Ch}}^t_{2k-1}$ and $\overline{\mathrm{Ch}}^s_{2k-1}$ are the transgression Chern characters in 4.7 with respect to the parameter t and s respectively, whereas $\overline{\overline{\mathrm{Ch}}}_{2k-2}(q_t^s)$ is a form in Ω^B_{2k-2}. Now, applying D to (4.12), we obtain that 0 on the left side is equal to

(4.13)

$$\begin{aligned} 0 &= D\,\mathrm{Ch}_{2k}(\tilde{q}) \\ &= \left(ds \wedge \frac{\partial}{\partial s} + dt \wedge \frac{\partial}{\partial t}\right) \mathrm{Ch}_{2k}(q_t^s) + d^B \left(\overline{\mathrm{Ch}}^t_{2k-1}(q_t^s) \wedge dt + \overline{\mathrm{Ch}}^s_{2k-1}(q_t^s) \wedge ds\right) \\ &\quad + \left(-\frac{\partial}{\partial s}\overline{\mathrm{Ch}}^t_{2k-1}(q_t^s) + \frac{\partial}{\partial t}\overline{\mathrm{Ch}}^s_{2k-1}(q_t^s) + d^B\overline{\overline{\mathrm{Ch}}}_{2k-2}(q_t^s)\right) \wedge dt \wedge ds. \end{aligned}$$

In particular, $\frac{\partial}{\partial s}\overline{\mathrm{Ch}}^t_{2k-1}(q_t^s) + \frac{\partial}{\partial t}\overline{\mathrm{Ch}}^s_{2k-1}(q_t^s) + d^B\overline{\overline{\mathrm{Ch}}}_{2k-2}(q_t^s) = 0$. So we have that

$$
\begin{aligned}
(4.14)\qquad & \int_0^1 \left(\overline{\mathrm{Ch}}^t_{2k-1}(q_t^0) - \overline{\mathrm{Ch}}^t_{2k-1}(q_t^1)\right) dt \\
&= \int_0^1 \int_0^1 \frac{\partial}{\partial s} \overline{\mathrm{Ch}}^t_{2k-1}(q_t^s) dt \wedge ds \\
&= - \int_0^1 \int_0^1 \left(\frac{\partial}{\partial t} \overline{\mathrm{Ch}}^s_{2k-1}(q_t^s) + d^B \overline{\overline{\mathrm{Ch}}}_{2k-1}(q_t^s)\right) dt \wedge ds \\
&= - \int_0^1 \left(\overline{\mathrm{Ch}}^s_{2k-1}(q_0^s) - \overline{\mathrm{Ch}}^s_{2k-1}(q_1^s)\right) ds + d^B \omega \\
&= -\Phi_{2k-1}\left(\int_0^1 \left(\overline{\mathrm{Ch}}^s_{2k-1}(p_0^s) - \overline{\mathrm{Ch}}^s_{2k-1}(p_1^s)\right) ds\right) + d^B \omega
\end{aligned}
$$

for some $\omega \in \Omega^B_{2k-1}$. This implies that

(4.15)

$$
\begin{aligned}
&\mathrm{Ch}^{rel}_{2k}([p_0^0, p_1^0; q_t^0]) - \mathrm{Ch}^{rel}_{2k}([p_0^1, p_1^1; q_t^1]) \\
&\quad = \Big((\mathrm{Ch}_{2k}(p_0^0) - \mathrm{Ch}_{2k}(p_0^1)) - (\mathrm{Ch}_{2k}(p_1^0)\, \mathrm{Ch}_{2k}(p_1^1)), \\
&\qquad\qquad \int_0^1 (\overline{\mathrm{Ch}}_{2k-1}(q_t^0) - \overline{\mathrm{Ch}}_{2k-1}(q_t^1)) dt\Big) \\
&\quad = \left(d^A \int_0^1 \left(\overline{\mathrm{Ch}}^s_{2k-1}(p_0^s) - \overline{\mathrm{Ch}}^s_{2k-1}(p_1^s)\right) ds,\right. \\
&\qquad\qquad \left. - \Phi_{2k-1}\left(\int_0^1 \left(\overline{\mathrm{Ch}}^s_{2k-1}(p_0^s) - \overline{\mathrm{Ch}}^s_{2k-1}(p_1^s)\right) ds\right) + d^B \omega\right)
\end{aligned}
$$

where we have applied (4.10) to the first entry. The last term is a boundary in the mapping cone complex. This, together with the fact that evidently Ch^{rel}_{2k} is unchanged under stabilizations, concludes the proof. □

PROPOSITION 4.15. *Let us consider the following commutative square of DGA morphisms*

$$
\begin{array}{ccc}
\Omega^A_\bullet & \xrightarrow{\Phi_\bullet} & \Omega^B_\bullet \\
\downarrow{\scriptstyle \alpha_\bullet} & & \downarrow{\scriptstyle \beta_\bullet} \\
\Omega^{A'}_\bullet & \xrightarrow{\Phi'_\bullet} & \Omega^{B'}_\bullet
\end{array}
$$

Let $(\alpha_0, \beta_0)_* \colon K_0(A \to B) \to K_0(A' \to B')$ *and* $(\alpha, \beta)_* \colon H_*(\Phi) \to H_*(\Phi')$ *denote the map induced in relative K-theory and relative non-commutative homology, respectively. Then we have that the relative Chern character is functorial, namely*

$$(\alpha, \beta)_* \mathrm{Ch}^{rel}_*([p_0, p_1; q_t]) = \mathrm{Ch}^{rel}_*((\alpha_0, \beta_0)_*[p_0, p_1; q_t]) \in H_*(\Phi').$$

PROOF. This is a straightforward consequence of the second part of Lemma 4.11. □

REMARK 4.16. Notice that the definition of $\overline{\mathrm{Ch}}(p_t)$ is coherent with formula [**49**, (1.46)], since up to commutators

$$\begin{aligned}\Xi_t\Theta^{k-1} &= (p_t dp_t \dot{p}_t + p_t\dot{p}_t dp_t)(p_t dp_t dp_t)^{k-1}\\ &= (p_t dp_t dp_t)^{k-1} p_t dp_t \dot{p}_t + p_t \dot{p}_t dp_t (p_t dp_t dp_t)^{k-1}\\ &= (p_t dp_t dp_t)^k \dot{p}_t + p_t\dot{p}_t(1-p_t)(dp_t dp_t)^{k-1} dp_t\\ &= (\dot{p}_t p_t + p_t \dot{p}_t(1-p_t))(dp_t dp_t)^{k-1} dp_t \qquad (4.16)\\ &= ((1-p_t)(p_t\dot{p}_t - \dot{p}_t) + p_t\dot{p}_t) dp_t (dp_t dp_t)^{k-1}\\ &= ((p_t-1)\dot{p}_t + p_t\dot{p}_t) dp_t (dp_t dp_t)^{k-1}\\ &= (2p_t - 1)\dot{p}_t dp_t (dp_t dp_t)^{k-1}\end{aligned}$$

which is exactly the integrand in [**49**, (1.46)].

Let us now consider the odd case. Let $[u]$ be a class in $K_1(A)$. By suspension, it corresponds to a class $[p_t] \in K_0(A\widehat{\otimes}C_0^\infty(0,1))$. Then we have the following definition.

DEFINITION 4.17. The degree $2k+1$ part of the Chern character of $[u] \in K_1(A)$ associated to the DGA $\Omega_\bullet^A$ over A is defined as the class

$$\mathrm{Ch}_{2k+1}([u]) := \int_0^1 \overline{\mathrm{Ch}}_{2k+1}(p_t)\, dt \in H_{2k+1}(\Omega_\bullet^A). \qquad (4.17)$$

Of course, there is also a direct description of $\mathrm{Ch}_{2k+1}([u])$ in terms of the invertible element u. We use suspension because it is more convenient for our treatment of relative K-theory. We record an alternative description in the following lemma, proved e.g. in [**62**, Proposition 1.170].

LEMMA 4.18. *Let $[u] \in K_1(A)$ be represented by an $n \times n$ invertible matrix u, and let $\Omega_\bullet^A$ be a DGA over A. Assume that u remains invertible in $\Omega_\bullet^A$. Then we have*

(4.18)
$$\mathrm{Ch}_{2k+1}([u]) = c_k \operatorname{Tr}(u^{-1}du(du^{-1}du)^k) = c_k \operatorname{Tr}((u^{-1}du)^{2k+1}) \in H_{2k+1}(\Omega_\bullet^A),$$

where c_k is a universal constant.

PROOF. Strictly speaking, [**62**] works with the cyclic bicomplex instead of non-commutative de Rham homology. The computations, however, translate directly. The universal constant can be derived from the proof of [**62**], which is a bit cumbersome because of the various normalizations chosen which one has to keep track of. □

Let now $[u; v_t]$ be a relative class in $K_1(\varphi)$; thus v_t is a path of invertible elements from $\varphi(u)$ to the identity. Then by suspension, it corresponds to a class $[p_t^0, p_t^1; q_t^s]$ in the relative K-theory of $\varphi \otimes \mathrm{id}\colon A\widehat{\otimes}C_0^\infty(0,1) \to B\widehat{\otimes}C_0^\infty(0,1)$, where these projectors are explicitly given in [**62**, p. 322].

DEFINITION 4.19. The degree $2k$ part of the relative Chern character of $[u, v_t] \in K_1(\varphi)$ associated to a pair of DGAs $\Phi_\bullet\colon \Omega_\bullet^A \to \Omega_\bullet^B$ over φ is defined as the relative class

$$\operatorname{Ch}^{rel}_{2k}([u;v_t]) := \Big(\int_0^1 \overline{\operatorname{Ch}}_{2k+1}(p^0_t)\, dt - \int_0^1 \overline{\operatorname{Ch}}_{2k+1}(p^1_t)\, dt, \\ \int_0^1 \int_0^1 \overline{\overline{\operatorname{Ch}}}_{2k}(q_{t,s}) \wedge ds \wedge dt \Big) \in H_{2k}(\Phi). \tag{4.19}$$

The fact that the odd absolute and relative Chern characters are well-defined and functorial is proved exactly as Proposition 4.14.

We note that relative and absolute Chern character are compatible to each other in the following sense:

PROPOSITION 4.20. *For an algebra homomorphism $\varphi\colon A \to B$ and a homomorphism of DGAs $\Phi\colon \Omega^A_\bullet \to \Omega^B_\bullet$ covering φ, the following diagram is commutative:*

$$\begin{array}{ccc} K_*(\varphi) & \longrightarrow & K_*(A) \\ \big\downarrow {\scriptstyle \operatorname{Ch}^{rel}_{2k-1+*}} & & \big\downarrow {\scriptstyle \operatorname{Ch}_{2k+*}} \\ H_{2k-1+*}(\Phi) & \xrightarrow{\ \partial\ } & H_{2k+*}(\Omega^A_\bullet) \end{array}$$

PROOF. This is a direct consequence of the definitions and holds on the cycle level: the natural map $K_*(\varphi) \to K_*(A)$ is given by projection onto the A-summand of a cycle, and the same applies to the boundary map $H_{2k-1+*}(\Phi) \to H_{2k+*}(\Omega^A_\bullet)$, and the A-summand of the relative Chern character is precisely the absolute A-Chern character. □

DEFINITION 4.21. We define the *total Chern character* $K_*(A) \to H_{[*]}(\Omega^A_\bullet)$ as the product of the Ch_{*+2k}. Here, we write $H_{[*]}(\Omega^A_\bullet) := \prod_{k\in\mathbb{Z}} H_{*+2k}(\Omega^A_\bullet)$, homology made $\mathbb{Z}/2$-periodic, and correspondingly for the relative Chern character $K_*(\varphi) \to H_{[*+1]}(\Phi)$.

Finally, we will need the following result.

LEMMA 4.22. *Let β be an idempotent matrix over the unitalization of $C_0((0,1)\times(0,1))$ representing the standard generator of $K_0(C_0((0,1)\times(0,1)))$. Then we have the identity*

$$\int_0^1 \int_0^1 \overline{\overline{\operatorname{Ch}}}_{2k-2}(p\otimes\beta) dt \wedge ds = \operatorname{Ch}_{2k}(p) \in H^+_{2k}(A) \qquad \textit{for all } [p] \in K_0(A).$$

PROOF. Choose $\beta \in C^\infty([0,1]\times[0,1])$ which is constant near the boundary. It is a standard fact that then $\int_0^1\int_0^1 \operatorname{Tr}(\beta d\beta d\beta) = 1$. Then one has the following calculations

$$\begin{aligned} \operatorname{Ch}_{2k}(p) &= \operatorname{Ch}_{2k}(p) \otimes 1 \\ &= \operatorname{Ch}_{2k}(p) \otimes \int_0^1\int_0^1 \operatorname{Tr}(\beta d\beta d\beta) \\ &= \int_0^1\int_0^1 \operatorname{Ch}_{2k}(p) \otimes \operatorname{Tr}(\beta d\beta d\beta) \\ &= \int_0^1\int_0^1 \operatorname{Ch}_{2k}(p\otimes\beta) \\ &= \int_0^1\int_0^1 \overline{\overline{\operatorname{Ch}}}_{2k-2}(p\otimes\beta) dt \wedge ds. \end{aligned}$$

In the last equality we used formula (4.12). For the last but one equality we observe that

$$\mathrm{Ch}_{2k}(p\otimes\beta) = \mathrm{Tr}(p\otimes\beta d(p\otimes\beta)d(p\otimes\beta))^k = \mathrm{Tr}(p\otimes\beta(dp\otimes\beta+p\otimes d\beta)^k);$$

thus the equality follows because all the terms in the integral which involve higher powers of $d\beta d\beta$ disappear. □

4.3. Excision in non-commutative de Rham homology

We have introduced relative de Rham cohomology for a DGA map which, by construction, appears as third term in an associated long exact sequence in non-commutative de Rham homology. Excision results concern the question under which conditions an absolute non-commutative de Rham homology can be used instead of the relative one, and we discuss some aspects of this briefly in this subsection.

Let

$$0 \to \Omega^I \xrightarrow{\iota} \Omega^A \xrightarrow{q} \Omega^Q \to 0 \tag{4.20}$$

be an extension of Frèchet DGAs. A natural question now is when the cohomology of Ω^I can be used instead of the (shifted) relative cohomology of q in an associated long exact sequence in homology.

A first guess might be that this should always be the case and just be based on general homological algebra. There is one subtlety, however: before taking homology of a chain complex, we pass to the quotient modulo commutators, and this process in general will not preserve exactness. The relevant notion is the following.

DEFINITION 4.23. The DGA Ω^I is called *split excisive* if $(\Omega^I)^2=\Omega^I$. In the Fréchet world we define that it is split excisive if $\overline{(\Omega^I)^2}=\Omega^I$.

LEMMA 4.24. *Assume that Ω^I is split excisive and that the short exact sequence* (4.20) *has a DGA split $\alpha\colon \Omega^Q\to\Omega^A$. Then we have a long exact sequence in non-commutative de Rham homology*

$$\cdots \to H_{*+1}(\Omega^Q) \xrightarrow{\partial} H_*(\Omega^I) \xrightarrow{\iota_*} H_*(\Omega^A) \xrightarrow{q_*} H_*(\Omega^Q) \xrightarrow{\delta} \cdots. \tag{4.21}$$

In general, define $H_*(\Omega^I\subset\Omega^A)$ as the homology of the chain complex $\Omega^I/(\Omega^I\cap \overline{[\Omega^A,\Omega^A]})$, i.e. we divide by all elements in Ω^I which are (limits of sums of) commutators of elements in Ω^A. As the differential preserves commutators and preserves Ω^I, this is indeed a chain complex. We get a long exact sequence in non-commutative de Rham homology

$$\cdots \to H_{*+1}(\Omega^Q) \xrightarrow{\partial} H_*(\Omega^I\subset\Omega^A) \xrightarrow{\iota_*} H_*(\Omega^A) \xrightarrow{q_*} H_*(\Omega^Q) \xrightarrow{\delta} \cdots. \tag{4.22}$$

There is a canonical map $H_*(\Omega^I)\to H_*(\Omega^I\subset\Omega^A)$ induced by the canonical projection of chain complexes defining the homology groups.

REMARK 4.25. Here, the notation $H_*(\Omega^I\subset\Omega^A)$ is *not* supposed to indicate a relative homology group, it is rather meant to be the corrected replacement of $H_*(\Omega^I)$ which is needed for excision, i.e. to obtain the exact sequence (4.22) as replacement of the desired (4.21) which holds only if Ω^I is excisive. The relative homology would be denoted $H_*(\Omega^I\to\Omega^A)$.

PROOF. The short exact sequence $0 \to \Omega^I \to \Omega^A \to \Omega^Q$ induces by elementary considerations an exact sequence

$$\Omega^I_{ab} \to \Omega^A_{ab} \to \Omega^Q_{ab} \to 0.$$

However, the first map will in general not be injective: precisely those elements in Ω^I which are commutators in Ω^A are mapped to zero. We obtain the desired short exact sequence of chain complexes

$$0 \to \Omega^I/(\Omega^I \cap \overline{[\Omega^A, \Omega^A]}) \to \Omega^A_{ab} \to \Omega^Q_{ab} \to 0$$

and homological algebra gives the long exact sequence (4.22).

For the split excisive case, we simply have to check that $\overline{[\Omega^I, \Omega^I]} = \Omega^I \cap \overline{[\Omega^A, \Omega^A]}$. Write $\Omega^{Q'} := \alpha(\Omega^Q) \subset \Omega^A$, this is a sub-DGA and $\Omega^A = \Omega^I + \Omega^{Q'}$ with $\Omega^I \cap \Omega^{Q'} = \{0\}$. Note that this is not a direct sum of DGAs in general, as we might have non-trivial mixed products, but it is a direct sum of Frèchet spaces.

We have

$$[\Omega^A, \Omega^A] = \underbrace{[\Omega^I, \Omega^I]}_{\subset \Omega^I} + \underbrace{[\Omega^I, \Omega^{Q'}]}_{\subset \Omega^I} + \underbrace{[\Omega^{Q'}, \Omega^{Q'}]}_{\subset \Omega^{Q'}}.$$

Therefore $\Omega^I \cap \overline{[\Omega^A, \Omega^A]} = \overline{[\Omega^I, \Omega^I]} + \overline{[\Omega^I, \Omega^A]}$ because $\Omega^I \cap \Omega^{Q'} = \{0\}$.

Finally, using the assumption that Ω^I is Fréchet split excisive, we can for $a \in \Omega^I$ and $T \in \Omega^A$ choose $u_i, v_i \in \Omega^I$ for $i \in \mathbb{N}$, with $a = \sum_{i \in \mathbb{N}} u_i v_i$ and obtain (in case u_i, v_i, T are all of even degree)

$$\begin{aligned} [a, T] &= \sum_{i \in \mathbb{N}} (u_i v_i T - T u_i v_i) \\ &= \sum_{i \in \mathbb{N}} \Big((u_i (v_i T) - (v_i T) u_i) + (v_i (T u_i) - (T u_i) v_i) \Big) \in \overline{[\Omega^I, \Omega^I]}. \end{aligned}$$

Also in the case of general degree, an easy computation shows that the signs for the graded commutator work out right and we get the desired inclusion $[\Omega^I, \Omega^A] \subset \overline{[\Omega^I, \Omega^I]}$ ($\overline{[\Omega^I, \Omega^A]} \subset \overline{[\Omega^I, \Omega^I]}$), resulting in the identification

$$\Omega_{I,ab} \xrightarrow{\cong} \Omega^I/(\Omega^I \cap \overline{[\Omega^A, \Omega^A]})$$

and therefore $H_*(\Omega^I) \xrightarrow{\cong} H_*(\Omega^I \subset \Omega^A)$. □

4.4. Splitting constructions in relative K-theory and homology

In this subsection, we discuss excision style results for relative K-theory and for de Rham cohomology in the special case of split short exact sequences. This will be needed in Section 6 for our construction of the Chern character map from analytic surgery to homology, i.e. for the main construction of this paper.

Let us fix the following algebraic setting: A is a Fréchet algebra, dense and holomorphically closed in a C*-algebra **A**; moreover let $I \subset A$ be a closed ideal and $B \subset A$ a closed subalgebra such that $I \cap B = \{0\}$.

REMARK 4.26. Because of the condition $I \cap B = \{0\}$ we have the split exact sequence (with obvious maps)

$$0 \longrightarrow I \longrightarrow I + B \underset{\alpha}{\overset{}{\rightleftarrows}} B \longrightarrow 0,$$

which gives an isomorphism $K_*(I) \cong K_*(I+B)/\alpha_*(K_*(B))$.

Let $\partial\colon K_*(B \to A/I) \to K_{*+1}(0 \to I)$ be the boundary map associated to the following short exact sequence

$$0 \longrightarrow (0 \to I) \longrightarrow (B \to A) \longrightarrow (B \to A/I) \longrightarrow 0\,,$$

where $(B \to A)$ and $(B \to A/I)$ denote the mapping cone algebras and $(0 \to I)$ is a way of writing the suspension of I as a mapping cone.

LEMMA 4.27. *The following diagram is commutative with vertical isomorphisms*
(4.23)

$$\begin{array}{ccccccccc}
\cdots K_{*-1}(I) & \xrightarrow{\iota_*\circ S} & K_*(B \to A) & \longrightarrow & K_*(B+I \to A) & \xrightarrow{\partial'} & K_*(I) & \xrightarrow{S} & \cdots \\
\Big\downarrow S & & \Big\downarrow \mathrm{id} & & q_* \Big\downarrow \cong & & \Big\downarrow S & & \\
\cdots K_*(0 \to I) & \xrightarrow{\iota_*} & K_*(B \to A) & \longrightarrow & K_*(B \to A/I) & \xrightarrow{\partial} & K_{*+1}(0 \to I) & \longrightarrow & \cdots
\end{array}$$

where $q\colon (B{+}I \to A) \to (B \to A/I)$ is the quotient by $(I \to I)$, S is the suspension isomorphism and ∂' is the following composition
(4.24)

$$K_*(B+I \to A) \longrightarrow K_*(B+I) \longrightarrow K_*(B+I)/\alpha_*(K_*(B)) \xleftarrow{\cong} K_*(I)\,.$$

Consequently the top row is a long exact sequence.

The top row could be interpreted as the long exact K-theory sequence of a hypothetical extension $0 \to (B \to A) \to (B+I \to A) \to (I \to 0) \to 0$. Of course, as B is *not* assumed to be an ideal in $B+I$, this is not an extension of C^*-algebras. Nonetheless, the lemma uses the split α to obtain the associated K-theory sequence, and gives explicitly the relevant maps.

PROOF. We only need to check the commutativity of the third square and this consists in chasing the following diagram.

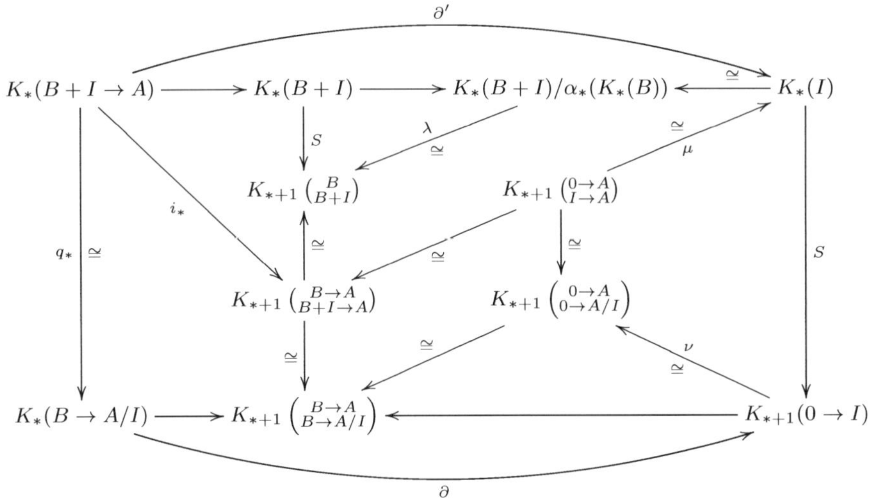

Here $K_{*+1}\begin{pmatrix} B \to A \\ B \to A/I \end{pmatrix}$ denotes the K-theory of the double mapping cone $((B \to A) \to (B \to A/I))$ (the same holds for the other similar K-groups). The map i_* is given by the composition of the suspension isomorphism and the natural inclusion; λ is induced by the obvious map $K_*(B+I) \to K_{*+1}(B \to B+I)$ and it is well defined on the quotient by $\alpha_* K_*(B)$ because its restriction to $\alpha_* K_*(B)$ factors through $K_{*+1}(B \to B) = 0$; μ denotes the quotient by $\begin{pmatrix} 0 \to A \\ 0 \to A \end{pmatrix}$ composed with the Bott periodicity isomorphism; ν is induced by the inclusion of I into A. The only part which is not clearly commutative is the right hand-side trapezoid, but this can be easily proved using [**90**, Lemma 3.6]. We leave the details to the reader, knowing that A, B and C in the notation of [**90**, Lemma 3.6] are here represented by $\begin{pmatrix} 0 \to A \\ I \to A \end{pmatrix}$, $\begin{pmatrix} 0 \to A \\ 0 \to A/I \end{pmatrix}$ and $(0 \to I)$ and that ∂_B is the isomorphism given by the composition of the Bott periodicity isomorphism and the inverse in K-theory of the inclusion of I in $(A \to A/I)$ and ∂_C is just the suspension isomorphism. $\square$

We now pass to the corresponding situation in non-commutative de Rham homology. For this, assume that Ω^A is a (Frèchet) DGA over A containing a closed DG ideal Ω^I over I and a closed sub-DGA Ω^B over B. Moreover, assume that $\Omega^I \cap \Omega^B = \{0\}$. Due to the fact that Ω^I is a DG ideal, $\Omega^I + \Omega^B$ is a DGA (over $I + B$). As for K-theory in Remark 4.26 the obvious inclusion defines a split α as DGA map of the sequence

$$0 \longrightarrow \Omega^I \longrightarrow \Omega^I + \Omega^B \underset{\curvearrowleft \alpha}{\longrightarrow} \Omega^B \longrightarrow 0\,. \tag{4.25}$$

For the following, we assume that Ω^I is split excisive as in Definition 4.23, which will be the case whenever we apply the general theory we are developing here.

REMARK 4.28. The following discussion easily generalizes to the case that Ω^I is not split excisive, just replacing $H_*(\Omega^I)$ by $H_*(\Omega^I \subset \Omega^I + \Omega^B)$ throughout. We decided to state it with the extra assumption because we will only apply this special case, and the notation then is a bit less cluttered.

PROPOSITION 4.29. *Adopt the situation described above. Let Ω^I be split excisive. The split α of (4.25) gives rise to a long exact sequence*

$$\cdots \to H_{[*-1]}(\Omega^I) \to H_{[*-1]}(\Omega^B \to \Omega^A) \to H_{[*-1]}(\Omega^I + \Omega^B \to \Omega^A) \xrightarrow{\delta'} H_{[*]}(\Omega^I) \to \cdots \tag{4.26}$$

where the boundary map $\delta' \colon H_{*-1}(\Omega^I + \Omega^B \to \Omega^A) \to H_*(\Omega^I)$ is defined by

$$H_{*-1}(\Omega^I + \Omega^B \to \Omega^A) \longrightarrow H_*(\Omega^B + \Omega^I) \longrightarrow H_*(\Omega^B + \Omega^I)/\alpha_*(H_*(\Omega^B)) \xleftarrow{\cong} H_*(\Omega^I). \tag{4.27}$$

PROOF. Since $(\Omega^I + \Omega^B)_{ab} = (\Omega^I)_{ab} + (\Omega^B)_{ab}$ by split excisiveness, we have the following exact sequence of complexes

$$0 \longrightarrow (\Omega^B \to \Omega^A)_{ab} \longrightarrow (\Omega^I + \Omega^B \to \Omega^A)_{ab} \xrightarrow{q} (\Omega^I \to 0)_{ab} \longrightarrow 0\,. \tag{4.28}$$

This gives rise to the long exact sequence

$$\cdots \longrightarrow H_*(\Omega^B \to \Omega^A) \longrightarrow H_*(\Omega^I + \Omega^B \to \Omega^A) \xrightarrow{q_*} H_*(\Omega^I \to 0) \xrightarrow{\delta} \cdots, \tag{4.29}$$

where δ is induced by the composition of the suspension $(\Omega^I \to 0)_\bullet \xrightarrow{S} (0 \to \Omega^I)_{\bullet+1}$ and the inclusion of $(0 \to \Omega^I)$ into $(\Omega^B \to \Omega^A)$.

Let us prove that the following square commutes.

$$\begin{array}{ccc} H_*(\Omega^I + \Omega^B \to \Omega^A) & \xrightarrow{q_*} & H_*(\Omega^I \to 0) \\ \downarrow{=} & & \downarrow{S} \\ H_*(\Omega^I + \Omega^B \to \Omega^A) & \xrightarrow{\delta'} & H_{*+1}(\Omega^I) \end{array}$$

Consider a class $[b+i, a] \in H_*(\Omega^I + \Omega^B \to \Omega^A)$, then by definition $\delta'[b+i, a] = [i] \in H_{*+1}(\Omega^I)$ which is exactly the suspension of $q_*[b+i, a] = [i, 0] \in H_*(\Omega^I \to 0)$. This proves the commutativity of the square and implies the exactness of (4.26). □

LEMMA 4.30. *In the situation described so far, the Chern character is compatible with the boundary maps ∂' of (4.24) and δ' of (4.27):*

$$\begin{array}{ccc} K_*(I + B \to A) & \xrightarrow{\partial'} & K_*(I) \\ \downarrow{\mathrm{Ch}^{rel}} & & \downarrow{\mathrm{Ch}} \\ H_{[*-1]}(\Omega^I + \Omega^B \to \Omega^A) & \xrightarrow{\delta'} & H_{[*]}(\Omega^I) \end{array}$$

PROOF. Thanks to Proposition 4.15, the Chern character is natural for maps of algebras covered by maps of DGAs and, by Proposition 4.20, for the boundary map in the usual pair sequence. We just have to apply this naturality to the individual maps whose composition defines ∂' and δ', see respectively (4.24) and (4.27). □

REMARK 4.31. It is clear that if the following square is commutative

$$\begin{array}{ccc} A & \xrightarrow{\alpha} & A' \\ \downarrow{\varphi} & & \downarrow{\varphi'} \\ B & \xrightarrow{\beta} & B' \end{array} \tag{4.30}$$

then from the definition of the relative Chern character we have that also the following square is commutative

$$\begin{array}{ccc} K_*(A \xrightarrow{\varphi} B) & \xrightarrow{(\alpha,\beta)_*} & K_*(A' \xrightarrow{\varphi'} B') \\ \downarrow{\mathrm{Ch}^{rel}} & & \downarrow{\mathrm{Ch}^{rel}} \\ H_*(A \to B) & \xrightarrow{(\alpha,\beta)_*} & H_*(A' \to B') \end{array}$$

namely the relative Chern character is functorial with respect to commutative squares as (4.30).

By the functoriality of the Chern character and its compatibility with the splitting given by α, we have the following result.

THEOREM 4.32. *Let Ω^I be split excisive. The following diagram is commutative*
(4.31)

$$\begin{array}{ccccccc}
K_{*+1}(I) & \xrightarrow{\iota_* \circ S} & K_*(B \to A) & \longrightarrow & K_*(B+I \to A) & \xrightarrow{\partial'} & K_*(I) \\
\downarrow \mathrm{Ch} & & \downarrow \mathrm{Ch}^{rel} & & \downarrow \mathrm{Ch}^{rel} & & \downarrow \mathrm{Ch} \\
H_{[*+1]}(\Omega^I) & \longrightarrow & H_{[*+1]}(\Omega^B \to \Omega^A) & \longrightarrow & H_{[*+1]}(\Omega^I + \Omega^B \to \Omega^A) & \xrightarrow{\delta'} & H_{[*]}(\Omega^I)
\end{array}$$

PROOF. For the left square, consider the diagram

$$\begin{array}{ccccc}
K_{*-1}(I) & \underset{\cong}{\xrightarrow{S}} & K_*(0 \to I) & \xrightarrow{\iota_*} & K_*(B \to A) \\
\downarrow \mathrm{Ch} & & \downarrow \mathrm{Ch}^{rel} & & \downarrow \mathrm{Ch}^{rel} \\
H_{[*+1]}(\Omega^I) & \xrightarrow{\equiv} & H_{[*+1]}(0 \to \Omega^I) & \xrightarrow{\iota_*} & H_{[*+1]}(\Omega^B \to \Omega^A)
\end{array} \tag{4.32}$$

For $*$ even, Ch: $K_1(I) \to H_{odd}(\Omega^I)$ is defined via the suspension isomorphism and $\overline{\mathrm{Ch}}_{odd}$, which also enters in the definition of Ch^{rel}. In this case, it is an immediate consequence of the definitions that the left square of (4.32) commutes. If $*$ is odd, by definition $\mathrm{Ch}^{rel} \circ S \colon K_0(I) \to H_{ev}(0 \to \Omega^I)$ is computed from the double suspension. Explicitly, if p is a projector over I, then

$$\mathrm{Ch}^{rel} \circ S(p) = [0, \int_0^1 \int_0^1 \overline{\overline{\mathrm{Ch}}}_{ev}(p \otimes \beta)\, dt \wedge ds] \in H_{\widehat{ev}}(0 \to \Omega^I),$$

where β is the Bott projector as in Lemma 4.22. By this lemma, this class equals $\mathrm{Ch}(p)$ under the identification $H_*(\Omega^I) \equiv H_*(0 \to \Omega^I)$. Therefore, the left square of (4.32) commutes in all degrees. The right square of (4.32) commutes because of naturality of the relative Chern character, Proposition 4.15. This also gives the commutativity of the middle square of (4.31). Moreover, (4.32) is a factorisation of the left square of (4.31) which therefore also commutes. Finally, the right square of (4.31) commutes because of Lemma 4.30. □

CHAPTER 5

Pseudodifferential operators and smooth subalgebras

The index theoretic approach to secondary invariants e.g. of metrics of positive scalar curvature requires the detailed work with operators derived from the Dirac operator. These live in various calculi of pseudodifferential operators and the properties of these calculi play an important role. For primary invariants, it turns out that the main aspects are captured by smoothing operators. However, this is different for the secondary invariants we want to study. Our setup leads us to deal with certain operators of order zero.

To capture the full information, in particular provided by the fundamental group, the efficient way to proceed is to enhance the bundles by twisting with the Mishchenko bundle and associated bundles. Equivalently, one can consider Γ-equivariant objects on the universal covering. This is perhaps less conceptual, but allows for explicit computations and fine tuned control due to the concreteness.

In this section we shall give a detailed account of classical facts about pseudodifferential calculus in the Mishchenko-Fomenko setting and holomorphically closed algebras of pseudodifferential operators, mostly following the work of John Lott [**51**].

5.1. Mishchenko bundle translated to universal covering

We want to work rather explicitly with the space of sections of the Mishchenko bundle and with certain classes of pseudodifferential operators on it. For this, it is useful and necessary to consider a couple of identifications. We use the following setup:

- a compact smooth manifold M without boundary and a complex vector bundle E over M;
- a Γ-principal bundle $\widetilde{M} \to M$, where Γ acts from the right and we denote by $R_\gamma \colon \widetilde{M} \to \widetilde{M}$ the (right) action map for $\gamma \in \Gamma$; this induces a natural action on $\widetilde{E}$, the equivariant lift of E to $\widetilde{M}$;
- dually, this action defines a right $\mathbb{C}\Gamma$-module structure on $C_c^\infty(\widetilde{M}, \widetilde{E})$ given by $f \cdot \gamma := R^*_{\gamma^{-1}} f$;
- the Mishchenko bundle

  $$\mathcal{V}_{\mathbb{C}\Gamma} := \widetilde{M} \times_\Gamma \mathbb{C}\Gamma$$

  is defined as the bundle associated to the right action of Γ on $\widetilde{M}$ and the left multiplication action of Γ on $\mathbb{C}\Gamma$: it is then defined as the quotient $\widetilde{M} \times \mathbb{C}\Gamma/\sim$, where $(R_{\gamma^{-1}}(\tilde{m}), \lambda) \sim (\tilde{m}, \gamma\lambda)$. This gives a flat bundle of $\mathbb{C}\Gamma$-right modules (fiberwise free of rank one) on M. More generally if Γ acts on an algebra $\mathcal{A}$ by right $\mathcal{A}$-module automorphisms (e.g. consider

$C^*_r\Gamma$ acted upon by left multiplication), we will denote by $\mathcal{V}_\mathcal{A}$ the bundle $\widetilde{M} \times_\Gamma \mathcal{A}$ of $\mathcal{A}$-right modules as before.

An important special case of this is when $\mathcal{A}$ is a DGA over $\mathbb{C}\Gamma$, e.g. $\Omega_\bullet(\mathbb{C}\Gamma)$. Then $\mathbb{C}\Gamma$-linear operators between $C^\infty(M; E \otimes \mathcal{V}_{\mathbb{C}\Gamma})$ and $C^\infty(M; E \otimes \mathcal{V}_\mathcal{A})$ extend uniquely to $\mathcal{A}$-linear operators on $C^\infty(M; E \otimes \mathcal{V}_\mathcal{A}) = C^\infty(M; E \otimes \mathcal{V}_{\mathbb{C}\Gamma} \otimes_{\mathbb{C}\Gamma} \mathcal{A})$. This is treated in detail in Section 5.2.

- let $\underline{\mathbb{C}\Gamma}$ be the product bundle $\widetilde{M} \times \mathbb{C}\Gamma$; Γ acts from the left on $C^\infty(\widetilde{M}; \widetilde{E} \otimes \underline{\mathbb{C}\Gamma})$ via $\gamma \cdot \sum_{g\in\Gamma} f_g\, g := \sum_{g\in\Gamma} R^*_\gamma f_g\, \gamma g$. This commutes with the right $\mathbb{C}\Gamma$-action which just multiplies on the right in the argument. Let $\mathcal{E}^{\mathrm{MF}}_{\mathbb{C}\Gamma}(\widetilde{M}; \widetilde{E})$ denote the Γ-invariant sections $C^\infty(\widetilde{M}; \widetilde{E} \otimes \underline{\mathbb{C}\Gamma})^\Gamma$, it is a standard fact that this right $\mathbb{C}\Gamma$-module identifies with $C^\infty(M; E \otimes \mathcal{V}_{\mathbb{C}\Gamma})$.

Lemma 5.1. *There is a canonical identification of $\mathbb{C}\Gamma$-modules*

$$L\colon C^\infty_c(\widetilde{M}; \widetilde{E}) \xrightarrow{\cong} C^\infty(\widetilde{M}; \widetilde{E} \otimes \underline{\mathbb{C}\Gamma})^\Gamma \xrightarrow{\cong} C^\infty(M; E \otimes \mathcal{V}_{\mathbb{C}\Gamma}).$$

The second isomorphism is a tautological one, just rewriting what a section of an associated bundle is in explicit terms. The first isomorphism is given by

$$s \mapsto \sum_{g\in\Gamma} (R^*_g s)\, g.$$

Proof. The proof of this lemma is exactly as the proof of [**51**, Proposition 5], but in fact easier since it deals with $\mathbb{C}\Gamma$ in place of the more complicated algebra $\mathcal{B}^\omega$, the algebra of function of exponential rapid decay. □

We shall now pass to pseudodifferential operators, concentrating first on the smoothing ones.

Definition 5.2. The smoothing operators $\Psi^{-\infty}_{\mathbb{C}\Gamma}(M, \mathcal{V}_{\mathbb{C}\Gamma})$ on the algebraic Mishchenko bundle $\mathcal{V}_{\mathbb{C}\Gamma}$ are the smooth sections of the bundle

$$\mathrm{Hom}_{\mathbb{C}\Gamma}(pr_2^*\mathcal{V}_{\mathbb{C}\Gamma}, pr_1^*\mathcal{V}_{\mathbb{C}\Gamma}) \to M \times M.$$

As kernel functions, they act in the usual way on the space of sections $C^\infty(M; \mathcal{V}_{\mathbb{C}\Gamma})$ as right $\mathbb{C}\Gamma$-linear operators.

It will be necessary for us to work very explicitly with this bundle and its sections and their action on sections of $\mathcal{V}_{\mathbb{C}\Gamma}$. This is based on the following lemma.

Lemma 5.3. *Let us consider* $\mathrm{End}_{\mathbb{C}\Gamma}(\mathbb{C}\Gamma)$ *where we use the right $\mathbb{C}\Gamma$-module structure on $\mathbb{C}\Gamma$. The bundle* $\mathrm{Hom}_{\mathbb{C}\Gamma}(pr_2^*\mathcal{V}_{\mathbb{C}\Gamma}, pr_1^*\mathcal{V}_{\mathbb{C}\Gamma})$ *over $M \times M$ is an associated bundle for the obvious right action of $\Gamma \times \Gamma$ on $\widetilde{M} \times \widetilde{M}$ and the left representation given by* $\Gamma \times \Gamma \to \mathrm{End}_{\mathbb{C}\Gamma}(\mathbb{C}\Gamma)$

$$(\gamma_1, \gamma_2) \cdot \phi = L_{\gamma_1} \circ \phi \circ L^{-1}_{\gamma_2}.$$

Recalling that $\mathrm{End}_{\mathbb{C}\Gamma}(\mathbb{C}\Gamma) \cong \mathbb{C}\Gamma$ *by means of the map $\phi \mapsto \phi(e)$, the action simply becomes*

$$(\gamma_1, \gamma_2) \cdot g = \gamma_1 g \gamma_2^{-1} \text{ for } g \in \mathbb{C}\Gamma.$$

In particular, we have the following identifications

$$\begin{aligned}\Psi^{-\infty}_{\mathbb{C}\Gamma}(M, \mathcal{V}_{\mathbb{C}\Gamma}) &\equiv C^\infty(M \times M; \mathrm{Hom}(pr_2^*\mathcal{V}, pr_1^*\mathcal{V}))\\ &\cong C^\infty(\widetilde{M} \times \widetilde{M}, \mathrm{End}_{\mathbb{C}\Gamma}(\mathbb{C}\Gamma))^{\Gamma\times\Gamma}\\ &\cong C^\infty(\widetilde{M} \times \widetilde{M}, \mathbb{C}\Gamma)^{\Gamma\times\Gamma}.\end{aligned}$$

If $k_1, k_2 \in C^\infty(\widetilde{M} \times \widetilde{M}, \mathbb{C}\Gamma)^{\Gamma\times\Gamma}$ *with* $k_i = \sum_{g\in\Gamma}(k_i)_g\, g$ *and* $f \in C^\infty(\widetilde{M}, \mathbb{C}\Gamma)^\Gamma$ *is of the form* $\sum_{g\in\Gamma} f_g\, g$ *(all sums are locally finite), the product and the action are given by*

$$(k_1 * k_2)_g(x,z) = \sum_{h\in\Gamma} \int_{\widetilde{M}/\Gamma} (k_1)_{gh^{-1}}(x,y)\cdot (k_2)_h(y,z)\, dy,$$

$$(k_1 f)_g(x) = \sum_{h\in\Gamma} \int_{\widetilde{M}/\Gamma} (k_1)_{gh^{-1}}(x,y) f_h(y)\, dy,$$

where we use that the expressions, as functions of y*, are* Γ*-invariant by the invariance properties we assume. Of course, the integral over* $\widetilde{M}/\Gamma$ *can also be written as an integral over a fundamental domain.*

PROOF. We observe that $\mathrm{Hom}_{\mathbb{C}\Gamma}(\mathcal{E},\mathcal{F}) = \mathcal{F}\otimes_{\mathbb{C}\Gamma}\mathcal{E}^*$ for two free finitely generated right $\mathbb{C}\Gamma$-modules, where $\mathcal{E}^* = \mathrm{Hom}_{\mathbb{C}\Gamma}(\mathcal{E},\mathbb{C}\Gamma)$. For us, $\mathcal{E} = \mathbb{C}\Gamma$ which has an additional Γ-action (by multiplication from the left), and $\gamma\in\Gamma$ acts on the left on $\mathcal{E}^*$ via $(\gamma\cdot f)(x) := f(\gamma^{-1}x)$ for $x\in\mathcal{E}$ and $f\in\mathcal{E}^*$. Under the identification $\mathbb{C}\Gamma \cong \mathrm{Hom}_{\mathbb{C}\Gamma}(\mathbb{C}\Gamma,\mathbb{C}\Gamma)$ this action becomes multiplication by γ^{-1} on the right and this finally gives the description of the associated bundle as claimed. The rest of the proposition follows from this description. □

Let us also fix the following identification of algebras

$$\Psi^{-\infty}_{\Gamma,c}(\widetilde{M}) \cong C^\infty_c(\widetilde{M}\times\widetilde{M})^\Gamma, \tag{5.1}$$

with composition on the left and convolution product on the right. It identifies a smoothing Γ-equivariant operator of Γ-compact support with its kernel, which is a smooth Γ-equivariant function on $\widetilde{M}\times\widetilde{M}$ with respect to the diagonal action of Γ and with Γ-compact support.

PROPOSITION 5.4. [**51**, Proposition 6] *We have the following isomorphism of *-algebras*

$$\begin{aligned} C^\infty_c(\widetilde{M}\times\widetilde{M})^\Gamma &\xrightarrow{\cong} C^\infty(\widetilde{M}\times\widetilde{M},\mathbb{C}\Gamma)^{\Gamma\times\Gamma} \\ k &\mapsto \sum_{\gamma\in\Gamma} R^*_{(\gamma,e)}k\,\gamma \end{aligned} \tag{5.2}$$

where $R^*_{g,h}k(x,y) := k(xg,yh)$*. Consequently*

$$\Psi^{-\infty}_{\Gamma,c}(\widetilde{M}) \overset{(5.1)}{\cong} C^\infty_c(\widetilde{M}\times\widetilde{M})^\Gamma \overset{(5.2)}{\cong} C^\infty(\widetilde{M}\times\widetilde{M},\mathbb{C}\Gamma)^{\Gamma\times\Gamma} \overset{5.3}{\cong} \Psi^{-\infty}_{\mathbb{C}\Gamma}(M,\mathcal{V}_{\mathbb{C}\Gamma}).$$

This composition is given by Ad_L*, the adjoint of the canonical isomorphism*

$$L\colon C^\infty_c(\widetilde{M}) \to C^\infty(\widetilde{M},\mathbb{C}\Gamma)^\Gamma.$$

All of this applies in a straightforward way if we add another auxiliary Γ-equivariant vector bundle $\widetilde{E}$, with algebra of smoothing operators $\Psi^{-\infty}_{\Gamma,c}(\widetilde{M},\widetilde{E})$, which is isomorphic to $\Psi^{-\infty}_{\mathbb{C}\Gamma}(M, E\otimes\mathcal{V}_{\mathbb{C}\Gamma})$.
Moreover, passing to completions, L clearly extends to an isomorphism of Hilbert $C^*_{red}\Gamma$-modules and the isomorphism (5.2) extends to an isomorphism of C*-algebras

$$Ad_L\colon C^*(\widetilde{M};\widetilde{E})^\Gamma \to \mathbb{K}(\mathcal{E}^{\mathrm{MF}}_{C^*_{red}\Gamma}(\widetilde{M},\widetilde{E})) \tag{5.3}$$

where $\mathcal{E}^{\mathrm{MF}}_{C^*_{red}\Gamma}(\widetilde{M};\widetilde{E})$ is the Hilbert-$C^*_{red}\Gamma$-module of the measurable sections of $E\otimes\mathcal{V}_{C^*_{red}\Gamma}$ with finite $C^*_{red}\Gamma$-valued norm, and $C^*(\widetilde{M};\widetilde{E})^\Gamma$ is the Roe algebra associated to the Γ-equivariant $C_0(\widetilde{M})$-module $L^2(\widetilde{M};\widetilde{E})$.

Let $\Psi^0_{\Gamma,c}(\widetilde{M},\widetilde{E})$ be the *-algebra of Γ-equivariant pseudodifferential operators of order 0 on $C^\infty(\widetilde{M};\widetilde{E})$ of Γ-compact support. We can extend Ad_L to an isomorphism between the multiplier algebra of $C^*(\widetilde{M};\widetilde{E}))^\Gamma$ and the multiplier algebra of $\mathbb{K}(\mathcal{E}^{\mathrm{MF}}_{C^*_{red}\Gamma}(\widetilde{M};\widetilde{E}))$, which is $\mathbb{B}(\mathcal{E}^{\mathrm{MF}}_{C^*_{red}\Gamma}(\widetilde{M};\widetilde{E}))$. This extension maps isomorphically the *-algebra $\Psi^0_{\mathbb{C}\Gamma}(\widetilde{M},\widetilde{E})$, which is certainly contained in the multiplier algebra of $C^*(\widetilde{M};\widetilde{E})^\Gamma$, to its image in $\mathbb{B}(\mathcal{E}^{\mathrm{MF}}_{C^*_{red}\Gamma}(\widetilde{M},\widetilde{E}))$.

Definition 5.5. Denote by $\Psi^0_{\mathbb{C}\Gamma}(\widetilde{M},\widetilde{E})$ the image of $\Psi^0_{\Gamma,c}(\widetilde{M},\widetilde{E})$ in

$$\mathbb{B}(\mathcal{E}^{\mathrm{MF}}_{C^*_{red}\Gamma}(\widetilde{M};\widetilde{E}))$$

through the extension of Ad_L.

Generalizing (5.2), elements in $\Psi^0_{\mathbb{C}\Gamma}(\widetilde{M},\widetilde{E})$ are locally finite sums of the form

$$\sum_{\gamma\in\Gamma} R^*_{(\gamma,e)}T(x,y)\,\gamma$$

where $T(x,y)$ is the pseudodifferential kernel of an operator in $\Psi^0_{\Gamma,c}(\widetilde{M},\widetilde{E})$ and γ acts on the $\mathbb{C}\Gamma$ part of the fiber by left multiplication; thus $R^*_{(\gamma,e)}T(x,y)$ has pseudodifferential singularities on the (γ,e)-translate of the diagonal of $\widetilde{M}\times\widetilde{M}$.

Remark 5.6. Note that the image of $\Psi^0_{\Gamma,c}(\widetilde{M},\widetilde{E})$ under Ad_L is a proper subalgebra of the Mishchenko-Fomenko algebra $\Psi^0_{\mathrm{MF}}(M,E\otimes\mathcal{V}_{\mathbb{C}\Gamma})$. This is why we use the notation involving $\widetilde{M}$: to stress that this remains pseudodifferential when interpreted as a lifted operator on the Γ-covering. For example, let us consider the multiplication by a central element $\gamma\in\Gamma$; this is certainly a pseudodifferential operator of order 0 in the Mishchenko-Fomenko calculus. An easy computation shows that this operator is the image under Ad_L of the translation operator $R^*_{\gamma^{-1}}$ which is not an element in $\Psi^0_{\mathbb{C}\Gamma}(\widetilde{M},\widetilde{E})$.

5.2. Pseudodifferential operators with coefficients in differential forms on $\mathbb{C}\Gamma$

All we did in the previous section can be extended to the situation where coefficients in non-commutative differential forms are taken into account. Recall $\Omega_\bullet(\mathbb{C}\Gamma)$, the universal differential algebra associated to $\mathbb{C}\Gamma$. We write e for the neutral element of Γ (the unit of $\mathbb{C}\Gamma$). As a vector space over $\mathbb{C}$, $\Omega_\bullet(\mathbb{C}\Gamma)$ has basis $\{g_0dg_1\dots dg_k\mid g_0,\dots,g_k\in\Gamma\}$ and the multiplication is given by the following formula

$$(g_0dg_1\dots dg_k)(g_{k+1}dg_{k+2}\dots dg_n)$$
$$:=\sum_{j=1}^{k}(-1)^{n-j}g_0dg_1\dots d(g_jg_{j+1})\dots dg_n+(-1)^n g_0g_1dg_2\dots dg_n,$$

while the differential is given by

$$d(g_0dg_1\dots dg_k)=dg_0dg_1\dots dg_k$$

First we fix some notation:

- if $\omega = g_1 \otimes \cdots \otimes g_n$, then we set $d\omega := dg_1 dg_2 \dots dg_n \in \Omega_\bullet(\mathbb{C}\Gamma)$ and $\pi(\omega) := g_1 g_2 \dots g_n \in \Gamma$.
- Observe that the elements of the form $\pi(\omega)^{-1}d\omega$ constitute a basis of the left $\mathbb{C}\Gamma$-module $\Omega_\bullet(\mathbb{C}\Gamma)$.
- Let $\mathcal{E}^{\mathrm{MF}}_{\Omega_\bullet(\mathbb{C}\Gamma)}(\widetilde{M};\widetilde{E})$ denote the right $\Omega_\bullet(\mathbb{C}\Gamma)$-module $C^\infty(\widetilde{M};\widetilde{E}\otimes\Omega_\bullet(\mathbb{C}\Gamma))^\Gamma$.

DEFINITION 5.7. Let us define a subalgebra $\Psi^0_{\Omega_\bullet(\mathbb{C}\Gamma)}(\widetilde{M},\widetilde{E})$ of the right $\Omega_\bullet(\mathbb{C}\Gamma)$-linear operators on the module $\mathcal{E}^{\mathrm{MF}}_{\Omega_\bullet(\mathbb{C}\Gamma)}(\widetilde{M};\widetilde{E})$ in the following way, extending Definition 5.5. Elements in $\Psi^0_{\Omega_\bullet(\mathbb{C}\Gamma)}(\widetilde{M},\widetilde{E})$ are given by sums that are locally finite over $\lambda \in \Gamma$ and finite over basic forms $\omega = g_1 \otimes \dots \otimes g_n$ of the following type

$$\mathbf{T} = \sum_{\lambda,\omega} R^*_{(\lambda,e)} T_\omega \otimes \lambda\pi(\omega)^{-1} d\omega$$

where T_ω are the distributional kernels of Γ-equivariant 0-order pseudodifferential operators on $\widetilde{M}$ of Γ-compact support and where $\lambda\pi(\omega)^{-1}d\omega$ acts by left multiplication in the $\Omega_\bullet(\mathbb{C}\Gamma)$ tensor factor in $C^\infty(\widetilde{M},\widetilde{E}\otimes\Omega_\bullet(\mathbb{C}\Gamma))^\Gamma$. The form degree in the expression for $\mathbf{T}$ defines a grading by non-negative integers which makes $\Psi^0_{\Omega_\bullet(\mathbb{C}\Gamma)}(\widetilde{M},\widetilde{E})$ a graded algebra. Its degree 0 part can be identified with $\Psi^0_{\mathbb{C}\Gamma}(\widetilde{M},\widetilde{E})$ as it follows from the expression for the elements in Definition 5.5. This is also explained by the identification of Remark 5.9 below.

Under the identifications in Section 5.1 of Γ-invariant objects on $\widetilde{M}$ and sections of bundles on M, $\Psi^0_{\Omega_\bullet(\mathbb{C}\Gamma)}(\widetilde{M},\widetilde{E})$ acts on the right $\Omega_\bullet(\mathbb{C}\Gamma)$-module of sections $C^\infty(M;E\otimes\mathcal{V}_{\Omega_\bullet(\mathbb{C}\Gamma)}) = \mathcal{E}^{\mathrm{MF}}_{\Omega_\bullet(\mathbb{C}\Gamma)}(\widetilde{M},\mathcal{E})$, as operators of order 0 in the Mishchenko-Fomenko calculus. But just as in the case without differential form coefficients in Remark 5.6, these are very special operators with constraints on the singularities, as explained in the following remark. This motivates our notation.

REMARK 5.8. Given an element $\mathbf{T}$ in $\Psi^0_{\Omega_\bullet(\mathbb{C}\Gamma)}(\widetilde{M},\widetilde{E})$, let us highlight that the only coefficients $R^*_{(\lambda,e)}T_\omega$ with pseudodifferential singularities on the diagonal of $\widetilde{M}\times\widetilde{M}$ are those with $\lambda = e$. Using Definition 6.1 in Section 6.1 below we thus see that the only coefficients $R^*_{(\lambda,e)}T_\omega$ with pseudodifferential singularities on the diagonal of $\widetilde{M}\times\widetilde{M}$ are those with $\lambda\pi(\omega)^{-1}d\omega \in \Omega^e_\bullet(\mathbb{C}\Gamma)$.

REMARK 5.9. The restriction to the sections of the degree 0 part of the forms gives a canonical bijection

$$\begin{aligned}(5.4)\quad &\mathrm{Hom}_{\Omega_\bullet(\mathbb{C}\Gamma)}(\mathcal{E}^{\mathrm{MF}}_{\Omega_\bullet(\mathbb{C}\Gamma)}(\widetilde{M},\widetilde{E}),\mathcal{E}^{\mathrm{MF}}_{\Omega_\bullet(\mathbb{C}\Gamma)}(\widetilde{M},\widetilde{E}))\\ &\qquad\qquad\xrightarrow{\cong}\mathrm{Hom}_{\mathbb{C}\Gamma}(\mathcal{E}^{\mathrm{MF}}_{\mathbb{C}\Gamma}(\widetilde{M},\widetilde{E}),\mathcal{E}^{\mathrm{MF}}_{\Omega_\bullet(\mathbb{C}\Gamma)}(\widetilde{M},\widetilde{E})),\end{aligned}$$

using that $\mathcal{E}^{\mathrm{MF}}_{\Omega_\bullet(\mathbb{C}\Gamma)}(\widetilde{M},\widetilde{E}) = C^\infty(M;E\otimes\mathcal{V}_{\mathbb{C}\Gamma}\otimes_{\mathbb{C}\Gamma}\Omega_\bullet(\mathbb{C}\Gamma)) = C^\infty(M;E\otimes\mathcal{V})\otimes_{\mathbb{C}\Gamma}\Omega_\bullet(\mathbb{C}\Gamma)$, so that the $\mathbb{C}\Gamma$-linear homomorphisms of the right hand side of (5.4) extend uniquely to $\Omega_\bullet(\mathbb{C}\Gamma)$-linear homomorphisms of the left hand side.

5.3. Mishchenko-Fomenko calculi associated to dense holomorphically closed subalgebras of $C^*_{red}\Gamma$

If now $\mathcal{A}\Gamma$ is a dense and holomorphically closed *-subalgebra of $C^*_{red}\Gamma$ which contains $\mathbb{C}\Gamma$, then $\Psi^{-\infty}_{\mathrm{MF}}(M, E \otimes \mathcal{V}_{\mathcal{A}\Gamma})$ is dense and holomorphically closed in

$$\Psi^{-\infty}_{\mathrm{MF}}(M, E \otimes \mathcal{V}_{C^*_{red}\Gamma}) \subset \mathbb{K}(\mathcal{E}^{\mathrm{MF}}_{C^*_{red}\Gamma}(\widetilde{M}, \widetilde{E})).$$

This follows from [**54**, Section 6]. Of course, this implies that $\Psi^{-\infty}_{\mathrm{MF}}(M, E \otimes \mathcal{V}_{\mathcal{A}\Gamma})$ is dense and holomorphically closed in $\mathbb{K}(\mathcal{E}^{\mathrm{MF}}_{C^*_{red}\Gamma}(\widetilde{M}, \widetilde{E}))$. Recall $\Psi^0_{\mathbb{C}\Gamma}(\widetilde{M}, \widetilde{E})$ and $\Psi^0_{\widehat{\Omega}_\bullet(\mathbb{C}\Gamma)}(\widetilde{M}, \widetilde{E})$ of Definitions 5.5 and 5.7.

DEFINITION 5.10. We define $\Psi^0_{\mathcal{A}\Gamma}(\widetilde{M}, \widetilde{E})$ as the *-subalgebra

$$\Psi^0_{\mathbb{C}\Gamma}(\widetilde{M}, \widetilde{E}) + \Psi^{-\infty}_{\mathrm{MF}}(M, E \otimes \mathcal{V}_{\mathcal{A}\Gamma}) \subset \Psi^0_{\mathrm{MF}}(M, E \otimes \mathcal{V}_{C^*_{red}\Gamma}).$$

Analogously, let $\Psi^0_{\widehat{\Omega}_\bullet(\mathcal{A}\Gamma)}(\widetilde{M}, \widetilde{E})$ denote the *-subalgebra

$$\Psi^0_{\widehat{\Omega}_\bullet(\mathbb{C}\Gamma)}(\widetilde{M}, \widetilde{E}) + \Psi^{-\infty}_{\mathrm{MF}}(M, E \otimes \mathcal{V}_{\widehat{\Omega}_\bullet(\mathcal{A}\Gamma)}) \subset \Psi^0_{\mathrm{MF}}(M, E \otimes \mathcal{V}_{\widehat{\Omega}_\bullet(C^*_{red}\Gamma)}).$$

REMARK 5.11. Recall that the notation $\widehat{\Omega}_\bullet$ for the universal differential algebra of a Fréchet algebra $\mathcal{A}$ means that we are using the projective tensor product in its definition, see for instance [**51**, Section 2].

PROPOSITION 5.12. *The *-subalgebra $Ad_L^{-1}(\Psi^0_{\mathcal{A}\Gamma}(\widetilde{M}, \widetilde{E}))$ is dense and holomorphically closed in $\Psi^0_\Gamma(\widetilde{M}, \widetilde{E})$, where we recall that $\Psi^0_\Gamma(\widetilde{M}, \widetilde{E})$ is the C^*-closure of $\Psi^0_{\Gamma,c}(\widetilde{M}, \widetilde{E})$. The corresponding result is true for operators with coefficients in forms.*

PROOF. Consider the algebras $\mathcal{J} := C^*(\widetilde{M}, L^2(\widetilde{M}, \widetilde{E}))^\Gamma$, $\mathcal{B} := \Psi^0_\Gamma(\widetilde{M}, \widetilde{E})$, $\mathcal{I} := Ad_L^{-1}(\Psi^{-\infty}(M, E \otimes \mathcal{V}_{\mathcal{A}\Gamma}))$ and $\mathcal{A} := Ad_L^{-1}(\Psi^0_{\mathcal{A}\Gamma}(\widetilde{M}, \widetilde{E}))$. All the hypotheses of [**43**, Theorem 4.2] are verified and the result follows. Notice that [**43**, Equation (4.4)] has a misprint. It should contain $\mathcal{A}$ instead of $\mathcal{B}$. □

CHAPTER 6

Mapping analytic surgery to homology

6.1. The delocalized homology exact sequence

"Delocalized" group (co)homology is an important theme for numeric secondary index invariants, but a bit subtle for the relevant case of completions of the group algebra. This is already treated in [**23**, Section 4.1]. Let us recall the basic definitions.

Let $\mathcal{A}\Gamma$ be a Fréchet *-algebra completion of $\mathbb{C}\Gamma$. The inclusion $\mathbb{C}\Gamma \hookrightarrow \mathcal{A}\Gamma$ induces the following morphism of DGAs (recall that $\widehat{\Omega}_\bullet$ is used for completions)

$$j\colon \Omega_\bullet(\mathbb{C}\Gamma) \to \widehat{\Omega}_\bullet(\mathcal{A}\Gamma)$$

and we will keep the same notation for the morphism induced between the abelianizations.

Definition 6.1. Let us denote by $\Omega^e_\bullet(\mathbb{C}\Gamma)$ the sub-DGA of $\Omega_\bullet(\mathbb{C}\Gamma)$ generated by those forms $g_0 dg_1 \dots dg_n$ such that $g_0 g_1 \dots g_n = e \in \Gamma$. Moreover, denote by $\Omega^{del}_\bullet(\mathbb{C}\Gamma)$ the complementary space to $\Omega^e_\bullet(\mathbb{C}\Gamma)$ in $\Omega_\bullet(\mathbb{C}\Gamma)$ spanned by $g_0 dg_1 \dots g_k$ with $g_0 g_1 \dots g_k \neq e \in \Gamma$.

We define the closed subcomplex $(\widehat{\Omega}^e_\bullet(\mathcal{A}\Gamma)_{ab}, d)$ of $(\widehat{\Omega}_\bullet(\mathcal{A}\Gamma)_{ab}, d)$ as the closure of $j(\Omega^e_\bullet(\mathbb{C}\Gamma)_{ab})$ and we denote the associated homology group by $H^e_*(\mathcal{A}\Gamma)$. Consider the resulting short exact sequence of complexes

$$0 \longrightarrow \widehat{\Omega}^e_\bullet(\mathcal{A}\Gamma)_{ab} \xrightarrow{\hat{j}} \widehat{\Omega}_\bullet(\mathcal{A}\Gamma)_{ab} \xrightarrow{q} \widehat{\Omega}^{del}_\bullet(\mathcal{A}\Gamma)_{ab} \longrightarrow 0 \tag{6.1}$$

where $\widehat{\Omega}^{del}_\bullet(\mathcal{A}\Gamma)_{ab} := \widehat{\Omega}_\bullet(\mathcal{A}\Gamma)_{ab}/\widehat{\Omega}^e_*(\mathcal{A}\Gamma)_{ab}$. We then have the associated long exact sequence of homology groups

(6.2)

$$\dots \longrightarrow H_*(\mathcal{A}\Gamma) \longrightarrow H^{del}_*(\mathcal{A}\Gamma) \xrightarrow{\delta_\Gamma} H^e_{*+1}(\mathcal{A}\Gamma) \longrightarrow H_{*-1}(\mathcal{A}\Gamma) \longrightarrow \dots$$

Whereas $\Omega^e_\bullet(\mathbb{C}\Gamma)_{ab}$ is a direct summand of $\Omega_\bullet(\mathbb{C}\Gamma)_{ab}$, it is not clear and in general false that the completion $\widehat{\Omega}^e_\bullet(\mathcal{A}\Gamma)_{ab}$ is a chain complex direct summand of $\widehat{\Omega}_\bullet(\mathcal{A}\Gamma)_{ab}$. Therefore, in general we can't expect that (6.2) splits into short exact sequences. To talk of the "localized" and "delocalized" parts of the homology is therefore not really appropriate (although frequently done).

6.2. Lott's connection and delocalized traces

Let $h\colon \widetilde{M} \to [0,1]$ be a cut-off function for the Γ-action on $\widetilde{M}$, namely a smooth function such that

$$\sum_{g\in\Gamma} R^*_g h(x) = 1 \quad \forall x \in \widetilde{M}. \tag{6.3}$$

Notice that, since the action of Γ is cocompact, h can be chosen with compact support.

DEFINITION 6.2. Lott's connection $\nabla^{\mathrm{Lott}}\colon \mathcal{E}^{\mathrm{MF}}_{\mathbb{C}\Gamma}(\widetilde{M},\widetilde{E}) \to \mathcal{E}^{\mathrm{MF}}_{\Omega_1(\mathbb{C}\Gamma)}(\widetilde{M},\widetilde{E})$ is defined by

$$\nabla^{\mathrm{Lott}}\left(\sum_{\lambda\in\Gamma} R^*_\lambda f\,\lambda\right) := \sum_{\lambda,g\in\Gamma} R^*_{\lambda g^{-1}}h\cdot R^*_\lambda f\otimes \lambda g^{-1}dg. \tag{6.4}$$

REMARK 6.3. Notice that ∇^{Lott} is an operator, but not a pseudodifferential operator: it involves pulling back with diffeomorphisms. Moreover, the map is not $\mathbb{C}\Gamma$ linear but acts as a connection for the multiplication by $\mathbb{C}\Gamma$.
We check that this definition is coherent with the original definition of Lott [**51**, Equation (41)], which is given by

$$\sum_{g\in\Gamma} hR^*_g f\otimes dg \qquad \text{for } f\in C^\infty_c(\widetilde{M},\widetilde{E}). \tag{6.5}$$

In Definition 6.2 we have written the expression (6.5) in our basis $\{g^{-1}dg\}$ of $\Omega_1(\mathbb{C}\Gamma)$ instead of the basis $\{dg\}$. Indeed

$$\sum_{g\in\Gamma} h\cdot R^*_g f\otimes dg = \sum_{g\in\Gamma} h\cdot R^*_g f\otimes gg^{-1}dg = \sum_{g\in\Gamma}(R^*_{g^{-1}}h)\cdot f\otimes g^{-1}dg,$$

finally, applying L of Lemma 5.1 we obtain (6.4). It is shown in [**51**, Proposition 9] that ∇^{Lott} is indeed a connection.

By standard arguments we can extend Lott's connection to $\mathcal{E}^{\mathrm{MF}}_{\Omega_\bullet(\mathbb{C}\Gamma)}(\widetilde{M},\widetilde{E})$ and we keep the notation ∇^{Lott} for the extended connection.

The curvature of this connection is given by

$$\Theta := (\nabla^{\mathrm{Lott}})^2 = \sum_{g_1,g_2} R^*_{(g_1g_2)^{-1}}h\cdot R^*_{g_2^{-1}}h\otimes (g_1g_2)^{-1}dg_1dg_2 \tag{6.6}$$

which is indeed pseudodifferential and $\mathbb{C}\Gamma$ linear. Therefore, by conjugation through L, it gives an element in $\mathrm{Hom}_{\mathbb{C}\Gamma}(\mathcal{E}^{\mathrm{MF}}_{\mathbb{C}\Gamma}(\widetilde{M},\widetilde{E}),\mathcal{E}^{\mathrm{MF}}_{\Omega_2(\mathbb{C}\Gamma)}(\widetilde{M},\widetilde{E}))$ which by Remark 5.9 we can also interpret as an element of $\mathrm{Hom}_{\Omega_\bullet(\mathbb{C}\Gamma)}(\mathcal{E}^{\mathrm{MF}}_{\Omega_\bullet(\mathbb{C}\Gamma)}(\widetilde{M},\widetilde{E}),\mathcal{E}^{\mathrm{MF}}_{\Omega_\bullet(\mathbb{C}\Gamma)}(\widetilde{M},\widetilde{E}))$.

LEMMA 6.4. *One has naturally a connection on $\Psi^0_{\Omega_\bullet(\mathbb{C}\Gamma)}(\widetilde{M},\widetilde{E})$ defined by*

$$\overline{\nabla}^{\mathrm{Lott}}(\mathbf{T}) = [\nabla^{\mathrm{Lott}},\mathbf{T}] := \nabla^{\mathrm{Lott}}\mathbf{T} - (-1)^{\deg\mathbf{T}}\mathbf{T}\nabla^{\mathrm{Lott}}. \tag{6.7}$$

PROOF. The connection property is obvious, as always if something is defined as a commutator with a fixed endomorphism: if $\mathbf{T},\mathbf{S}\in\Psi^0_{\Omega_\bullet(\mathbb{C}\Gamma)}(\widetilde{M},\widetilde{E})$ then

$$\begin{aligned}\overline{\nabla}^{\mathrm{Lott}}(\mathbf{TS}) &= \nabla^{\mathrm{Lott}}\mathbf{TS} - (-1)^{\deg(\mathbf{TS})}\mathbf{TS}\nabla^{\mathrm{Lott}}\\ &= \nabla^{\mathrm{Lott}}\mathbf{TS} - (-1)^{\deg(\mathbf{T})}\mathbf{T}\nabla^{\mathrm{Lott}}\mathbf{S} + (-1)^{\deg(\mathbf{T})}\mathbf{T}\nabla^{\mathrm{Lott}}\mathbf{S}\\ &\quad - (-1)^{\deg(\mathbf{T})}(-1)^{\deg(\mathbf{S})}\mathbf{TS}\nabla^{\mathrm{Lott}}\\ &= (\overline{\nabla}^{\mathrm{Lott}}(\mathbf{T}))\mathbf{S} + (-1)^{\deg(\mathbf{T})}\mathbf{T}(\overline{\nabla}^{\mathrm{Lott}}(\mathbf{S})).\end{aligned}$$

Next we have to prove that indeed $\overline{\nabla}^{\mathrm{Lott}}(T)\in\Psi^0_{\Omega_\bullet(\mathbb{C}\Gamma)}(\widetilde{M},\widetilde{E})$, which is not obvious. First, observe that the $\Omega_\bullet(\mathbb{C}\Gamma)$-linearity of $\mathbf{T}$ and the connection property of ∇ imply that $\overline{\nabla}(\mathbf{T})$ is again $\Omega_\bullet(\mathbb{C}\Gamma)$-linear:

$$\begin{aligned}\overline{\nabla}(\mathbf{T})(f\alpha) &= \nabla(\mathbf{T}(f\alpha)) - (-1)^{\deg(\mathbf{T})}\mathbf{T}(\nabla(f\alpha))\\ &= \nabla(\mathbf{T}(f)\alpha) - (-1)^{\deg(\mathbf{T})}(\mathbf{T}(\nabla(f)\alpha) + (-1)^{\deg(f)}\mathbf{T}(f d\alpha))\\ &= \nabla(\mathbf{T}(f))\alpha + (-1)^{\deg(\mathrm{T})+\deg(f)}\mathbf{T} f d\alpha - (-1)^{\deg(\mathbf{T})}\mathbf{T}(\nabla(f))\alpha\\ &= (-1)^{\deg(\mathbf{T})+\deg(f)}\mathbf{T}(f)d\alpha\\ &= \overline{\nabla}(\mathbf{T})(f)\alpha.\end{aligned}$$

To show that $\overline{\nabla}(\mathbf{T})$ indeed is pseudodifferential, let us follow Definition 5.7 and explicitly take

$$\mathbf{T} = \sum_{\gamma\in\Gamma,\omega} R^*_{(\gamma,e)}T_\omega(x,y)\gamma\pi(\omega)^{-1}d\omega,$$

with $\deg(\mathbf{T}) = \deg(\omega)$. Moreover, fix $f = \sum_{\lambda\in\Gamma} R^*_\lambda f\lambda \in \mathcal{E}^{\mathrm{MF}}_{\Omega_\bullet(\mathbb{C}\Gamma)}(\widetilde{M},\widetilde{E})$ of form degree 0 (this determines the general case by $\Omega_\bullet(\mathbb{C}\Gamma)$-linearity). Then

(6.8)

$$\begin{aligned}&\overline{\nabla}^{\mathrm{Lott}}(\mathbf{T})(f)\\ &= \nabla(\mathbf{T})(f) - (-1)^{\deg(\omega)}\mathbf{T}(\nabla f)\\ &= \nabla\left(\sum_{\gamma,\omega,\lambda}\int_{\widetilde{M}/\Gamma} dy\; R^*_{\gamma,e}\mathbf{T}_\omega(x,y)R^*_\lambda f(y)\gamma\pi(\omega)^{-1}d\omega\cdot\lambda\right)\\ &\quad - (-1)^{\deg(\omega)}\sum_{\gamma,\omega}\int_{\widetilde{M}/\Gamma} dy\; R^*_{(\gamma,e)}\mathbf{T}_\omega(x,y)\gamma\pi(\omega)^{-1}d\omega\\ &\qquad\times\left(\sum_{\lambda,g} R^*_{\lambda g^{-1}}h(y)R^*_\lambda f(y)\lambda g^{-1}dg\right)\\ &= \sum_{g,\gamma,\omega,\lambda}\int_{\widetilde{M}/\Gamma} dy\; R^*_{\gamma g^{-1}}h(x)R^*_{\gamma,e}\mathbf{T}_\omega(x,y)R^*_\lambda f(y)\gamma g^{-1}dg\cdot\pi(\omega)^{-1}d\omega\cdot\lambda\\ &\quad + \sum_{\gamma,\omega,\lambda}\int_{\widetilde{M}/\Gamma} dy\; R^*_{\gamma,e}\mathbf{T}_\omega(x,y)R^*_\lambda f(y)\gamma\cdot d(\pi(\omega)^{-1}d\omega\cdot\lambda)\\ &\quad - (-1)^{\deg(\omega)}\sum_{\lambda,g,\gamma,\omega}\int_{\widetilde{M}/\Gamma} dy\; R^*_{(\gamma,e)}\mathbf{T}_\omega(x,y)R^*_{\lambda g^{-1}}h(y)R^*_\lambda f(y)\gamma\pi(\omega)^{-1}d\omega\lambda g^{-1}dg\\ &\overset{\tilde{g}=g\lambda^{-1}}{=} \sum_{g,\gamma,\omega,\lambda}\int_{\widetilde{M}/\Gamma} dy\; R^*_{\gamma g^{-1}}h(x)R^*_{\gamma,e}\mathbf{T}_\omega(x,y)R^*_\lambda f(y)\gamma g^{-1}dg\cdot\pi(\omega)^{-1}d\omega\cdot\lambda\\ &\quad + \sum_{\gamma,\omega,\lambda}\int_{\widetilde{M}/\Gamma} dy\; R^*_{\gamma,e}\mathbf{T}_\omega(x,y)R^*_\lambda f(y)\gamma\cdot d(\pi(\omega)^{-1})d\omega\cdot\lambda\\ &\quad + (-1)^{\deg(\omega)}\sum_{\gamma,\omega,\lambda}\int_{\widetilde{M}/\Gamma} dy\; R^*_{\gamma,e}\mathbf{T}_\omega(x,y)R^*_\lambda f(y)\gamma\cdot\pi(\omega)^{-1}d\omega\cdot d\lambda)\\ &\quad - (-1)^{\deg(\omega)}\sum_{\lambda,\tilde{g},\gamma,\omega}\int_{\widetilde{M}/\Gamma} dy\; R^*_{(\gamma,e)}\mathbf{T}_\omega(x,y)R^*_{\tilde{g}^{-1}}h(y)R^*_\lambda f(y)\gamma\pi(\omega)^{-1}d\omega\tilde{g}^{-1}d(\tilde{g}\lambda)\\ &= \sum_{g,\gamma,\omega,\lambda}\int_{\widetilde{M}/\Gamma} dy\; \gamma R^*_{\gamma,e}(R^*_{g^{-1}}h(x)g^{-1}dg\mathbf{T}_\omega(x,y)\pi(\omega)^{-1}d\omega)\cdot R^*_\lambda f(y)\lambda\end{aligned}$$

$$
\begin{aligned}
&+\sum_{\gamma,\omega,\lambda}\int_{\widetilde{M}/\Gamma} dy\ \gamma R^*_{\gamma,e}\mathbf{T}_\omega(x,y)\cdot d(\pi(\omega)^{-1})d\omega\cdot R^*_\lambda f(y)\lambda\\
&+(-1)^{\deg(\omega)}\sum_{\gamma,\omega,\lambda}\int_{\widetilde{M}/\Gamma} dy\ R^*_{\gamma,e}\mathbf{T}_\omega(x,y)R^*_\lambda f(y)\gamma\cdot\pi(\omega)^{-1}d\omega\cdot d\lambda\\
&-(-1)^{\deg(\omega)}\sum_{\lambda,g,\gamma,\omega}\int_{\widetilde{M}/\Gamma} dy\gamma R^*_{(\gamma,e)}\mathbf{T}_\omega(x,y)\pi(\omega)^{-1}d\omega R^*_{g^{-1}}h(y)g^{-1}dg\cdot R^*_\lambda f(y)\lambda\\
&-(-1)^{\deg(\omega)}\sum_{\lambda,\gamma,\omega}\int_{\widetilde{M}/\Gamma} dy R^*_{(\gamma,e)}\mathbf{T}_\omega(x,y)R^*_\lambda f(y)\gamma\cdot\pi(\omega)^{-1}d\omega\underbrace{\sum_g R^*_{g^{-1}}h(y)g^{-1}g}_{=1}\cdot d\lambda\\
&=\sum_{\gamma,\omega,\lambda}\int_{\widetilde{M}/\Gamma} dy\ \gamma R^*_{\gamma,e}\left([\sum_g R^*_{g^{-1}}h(x)g^{-1}dg,\mathbf{T}_\omega(x,y)\pi(\omega)^{-1}d\omega]_s+\mathbf{T}_\omega(x,y)\underbrace{\pi(\tilde{\omega})}_{=1}d\tilde{\omega}\right)\\
&\qquad\cdot R^*_\lambda f(y)\lambda
\end{aligned}
$$

where the last equality uses $\tilde{\omega}=(\pi(\omega)^{-1},\omega)$. Here, if $\omega=(g_1,\dots,g_{\deg(\omega)})$ we write $\tilde{\omega}:=((g_1\cdots g_{\deg(\omega)})^{-1},g_0,\dots,g_{\deg(\omega)})$, therefore $\pi(\tilde{\omega})=1$. We also write $[\cdot,\cdot]_s$ for the supercommutator, here with respect to the form degree.

From the explicit expression we immediately observe that $\overline{\nabla}(\mathbf{T})$ belongs to $\Psi^0_{\Omega_\bullet(\mathbb{C}\Gamma)}(\widetilde{M},\widetilde{E})$, i.e. is pseudodifferential with the right invariance properties. Note that, if the form degree of $\mathbf{T}$ is 0 the expression simplifies and the main term is the commutator of $\mathbf{T}$ and the multiplication operators $R^*_{g^{-1}}h$, which is what is obtained in [**51**, Equation (53)] for a special operator $\mathbf{T}$ (actually for a Dirac type operator which has positive degree, but the whole construction extends to this case anyway, compare Proposition 6.16). Notice that Lott's sign conventions must be modified for right modules, as done in [**44**]. □

It is immediate to see $(\overline{\nabla}^{\text{Lott}})^2(\mathbf{T})=\Theta\mathbf{T}-\mathbf{T}\Theta$ and that $\overline{\nabla}^{\text{Lott}}(\Theta)=0$.

Remark 6.5. Let now $\mathcal{A}\Gamma$ be any Fréchet completion of $\mathbb{C}\Gamma$ contained in $C^*_{red}\Gamma$. The proof of [**51**, Proposition 9] applies using $\mathcal{A}\Gamma$, therefore ∇^{Lott} is a well-defined connection on $\mathcal{E}^{\text{MF}}_{\mathcal{A}\Gamma}(\widetilde{M},\widetilde{E})$ and everything we have seen so far also holds in this setting.

Definition 6.6. In [**51**, Section II], Lott defines a $\widehat{\Omega}_\bullet(\mathcal{A}\Gamma)_{ab}$-valued trace

$$
\begin{aligned}
&\text{TR}\colon \Psi^{-\infty}_{\widehat{\Omega}_\bullet(\mathcal{A}\Gamma)}(\widetilde{M},\widetilde{E})\to\widehat{\Omega}_\bullet(\mathcal{A}\Gamma)_{ab};\\
&\qquad\qquad \mathbf{T}\mapsto\sum_{\lambda,\omega}\int_{\mathcal{F}}\text{Tr}(R^*_{(\lambda,e)}T_\omega(\tilde{x},\tilde{x}))d\text{vol}(\tilde{x})\,\lambda\pi(\omega)^{-1}d\omega,
\end{aligned}
\tag{6.9}
$$

where $\mathbf{T}=\sum_{\lambda,\omega}R^*_{(\lambda,e)}T_\omega\otimes\lambda\pi(\omega)^{-1}d\omega$ with T_ω smoothing Γ-equivariant kernels on $\widetilde{M}$. Here, $\mathcal{F}$ is a fundamental domain for the action of Γ on $\widetilde{M}$ and Tr is the fiberwise trace of vector bundle homomorphisms.

In this section, we will define in a similar fashion a delocalized trace on

$$\Psi^0_{\widehat{\Omega}_\bullet(\mathcal{A}\Gamma)}(\widetilde{M},\widetilde{E}).$$

Recall from Section 6.1 that, as a vector space, $\Omega_\bullet(\mathbb{C}\Gamma)$ decomposes into the direct sum $\Omega^e_\bullet(\mathbb{C}\Gamma)\oplus\Omega^{del}_\bullet(\mathbb{C}\Gamma)$. In turn, always as a vector space, Definition 5.7

shows immediately that $\Psi^0_{\Omega_\bullet(\mathbb{C}\Gamma)}(\widetilde{M},\widetilde{E})$ splits into the direct sum $\Psi^0_{\Omega^e_\bullet(\mathbb{C}\Gamma)}(\widetilde{M},\widetilde{E})\oplus\Psi^0_{\Omega^{del}_\bullet(\mathbb{C}\Gamma)}(\widetilde{M},\widetilde{E})$. Then we have a projection

$$\pi_{del}\colon \Psi^0_{\Omega_\bullet(\mathbb{C}\Gamma)}(\widetilde{M},\widetilde{E})\to\Psi^0_{\Omega^{del}_\bullet(\mathbb{C}\Gamma)}(\widetilde{M},\widetilde{E}).$$

Recall that, thanks to Remark 5.8, if we consider an element in the image of π_{del} then the restriction to the diagonal of its kernel is smooth. Hence we can give the following definition.

DEFINITION 6.7. Let $\mathbf{T}$ be an element of $\Psi^0_{\Omega_*(\mathbb{C}\Gamma)}(\widetilde{M},\widetilde{E})$. Define

$$\mathrm{TR}^{del}_0(\mathbf{T}):=\int_{\tilde{x}\in\mathcal{F}}\mathrm{Tr}\left(\pi_{del}(\mathbf{T})(\widetilde{x},\widetilde{x})\right)d\mathrm{vol}(\widetilde{x})\in\Omega^{del}_\bullet(\mathbb{C}\Gamma)_{ab},$$

where $\mathcal{F}$ is a fundamental domain for the action of Γ on $\widetilde{M}$.

Unrolling the definition, we obtain

$$\mathrm{TR}^{del}_0(\mathbf{T})=\sum_{\lambda\neq e,\omega}\int_{\mathcal{F}}\mathrm{Tr}(R^*_{(\lambda,e)}T_\omega(\tilde{x},\tilde{x}))d\mathrm{vol}(\tilde{x})\,\lambda\pi(\omega)^{-1}d\omega, \tag{6.10}$$

where it is important to notice that the sum is finite because all operators are given by locally finite sums. Furthermore, as for TR, it is straightforward to see that TR^{del}_0 is a trace and that Definition 6.7 does not depend on the choice of the fundamental domain $\mathcal{F}$.

We now want to extend the definition of the delocalized trace to $\Psi^0_{\widehat{\Omega}_\bullet(\mathcal{A}\Gamma)}(\widetilde{M},\widetilde{E})$. To this end, let us consider a smooth function χ on $\widetilde{M}\times\widetilde{M}$ such that it is equal to 1 in a neighborhood of the diagonal and is properly supported, namely the projections $\pi_1,\pi_2\colon\widetilde{M}\times\widetilde{M}\to\widetilde{M}$ are proper on the support of χ. Then an element $\mathbf{T}=\sum_{\lambda,\omega}R^*_{(\lambda,e)}T_\omega\otimes\lambda\pi(\omega)^{-1}d\omega\in\Psi^0_{\widehat{\Omega}_\bullet(\mathcal{A}\Gamma)}(\widetilde{M},\widetilde{E})$ decomposes as $\mathbf{T}=\mathbf{T}_0+\mathbf{T}_\infty$, where

$$\mathbf{T}_0=\sum_{\lambda,\omega}R^*_{(\lambda,e)}(\chi T_\omega)\otimes\lambda\pi(\omega)^{-1}d\omega,\quad \mathbf{T}_\infty=\mathbf{T}-\mathbf{T}_0.$$

Observe that by Definition 5.7, since χT_ω is Γ-compactly supported,

$$\mathbf{T}_0\in\Psi^0_{\Omega_\bullet(\mathbb{C}\Gamma)}(\widetilde{M},\widetilde{E})\quad\text{and}\quad\mathbf{T}_\infty\in\Psi^{-\infty}_{\widehat{\Omega}_\bullet(\mathcal{A}\Gamma)}(\widetilde{M},\widetilde{E}).$$

DEFINITION 6.8. We define $\mathrm{TR}^{del}\colon\Psi^0_{\widehat{\Omega}_\bullet(\mathcal{A}\Gamma)}(\widetilde{M},\widetilde{E})\to\widehat{\Omega}^{del}_\bullet(\mathcal{A}\Gamma)_{ab}$ for $\mathbf{T}=\mathbf{T}_0+\mathbf{T}_\infty$ decomposed as above by

$$\mathrm{TR}^{del}(\mathbf{T}):=\mathrm{TR}^{del}_0(\mathbf{T}_0)+\mathrm{TR}(\mathbf{T}_\infty).$$

REMARK 6.9. Let us prove that Definition 6.8 is independent of the choice of the cutoff function χ. Indeed, assume χ' is a second cutoff function and let $\mathbf{T}=\mathbf{T}'_0+\mathbf{T}'_\infty$ be the corresponding decomposition. Then $\mathbf{T}_0-\mathbf{T}'_0$ vanishes in a neighborhood of the diagonal and therefore $\mathbf{T}_0-\mathbf{T}'_0\in\Psi^{-\infty}_{\widehat{\Omega}^+_\bullet(\mathcal{A}\Gamma)}(\widetilde{M},\widetilde{E})$ and the summand of $\lambda=e$ in formula (6.9) for $\mathrm{TR}(\mathbf{T}_0-\mathbf{T}'_0)$ is zero, so that with Equation (6.10) $\mathrm{TR}(\mathbf{T}_0-\mathbf{T}'_0)=\mathrm{TR}^{del}_0(\mathbf{T}_0-\mathbf{T}'_0)$ and finally,

$$\begin{aligned}\mathrm{TR}^{del}_0(\mathbf{T}_0)+\mathrm{TR}(\mathbf{T}_\infty)-\mathrm{TR}^{del}_0(\mathbf{T}'_0)-\mathrm{TR}(\mathbf{T}'_\infty)&=\mathrm{TR}^{del}_0(\mathbf{T}_0-\mathbf{T}'_0)+\mathrm{TR}(\mathbf{T}_\infty-\mathbf{T}'_\infty)\\&=\mathrm{TR}(\mathbf{T}_0-\mathbf{T}'_0+\mathbf{T}_\infty-\mathbf{T}'_\infty)\\&=\mathrm{Tr}(0)=0.\end{aligned}$$

Moreover, notice that TR^{del} is a well-defined map with values in $\widehat{\Omega}^{del}_{\bullet}(\mathcal{A}\Gamma)$, but it is a trace only if we pass to the quotient $\widehat{\Omega}^{del}_{\bullet}(\mathcal{A}\Gamma)_{ab}$.

PROPOSITION 6.10. *If* $\mathbf{T} \in \Psi^0_{\Omega_\bullet(\mathbb{C}\Gamma)}(\widetilde{M}, \widetilde{E})$, *then*

$$d\mathrm{TR}^{del}_0(\mathbf{T}) = \mathrm{TR}^{del}_0([\nabla^{\mathrm{Lott}}, \mathbf{T}]) \in \Omega^{del}_{\bullet}(\mathbb{C}\Gamma)_{ab}.$$

PROOF. Let $\mathbf{T}$ be given by the sum $\sum_{\omega,\lambda} R^*_{(\lambda,e)} T_\omega \lambda\pi(\omega)^{-1}d\omega$. The direct calculation of (6.8) gives that $[\nabla^{\mathrm{Lott}}, \mathbf{T}]$ is equal to

$$\sum_{\lambda,\omega,\gamma} R^*_{\lambda\gamma^{-1}} h\, R^*_{(\lambda,e)} T_\omega\, \lambda\gamma^{-1} d\gamma \cdot \pi(\omega)^{-1} d\omega$$
$$+ \sum_{\lambda,\omega} R^*_{(\lambda,e)} T_\omega\, \lambda d\pi(\omega)^{-1} d\omega - \sum_{\lambda,\omega,\gamma} (-1)^{deg(d\omega)} R^*_{(\lambda,e)} T_\omega R^*_{\gamma^{-1}} h\, \lambda\pi(\omega)^{-1} d\omega \cdot \gamma^{-1} d\gamma.$$

Passing to the Mishchenko-Fomenko context and applying TR^{del} we obtain

$$\begin{aligned}
&\mathrm{TR}^{del}_0([\nabla^{\mathrm{Lott}}, \mathbf{T}]) \\
&\quad = \int_{x\in\mathcal{F}} \sum_{\lambda\neq e,\omega,\gamma} \mathrm{Tr}\left(R^*_{\lambda\gamma^{-1}} h(x)\, R^*_{(\lambda,e)} T_\omega(x,x)\right) \lambda\gamma^{-1} d\gamma \cdot \pi(\omega)^{-1} d\omega \\
&\qquad + \int_{x\in\mathcal{F}} \sum_{\lambda\neq e,\omega} \mathrm{Tr}\left(R^*_{(\lambda,e)} T_\omega(x,x)\right) \lambda d\pi(\omega)^{-1} d\omega \\
&\qquad - \int_{x\in\mathcal{F}} \sum_{\lambda\neq e,\omega,\gamma} (-1)^{deg(d\omega)}\, \mathrm{Tr}\left(R^*_{(\lambda,e)} T_\omega(x,x) R^*_{\gamma^{-1}} h(x)\right) \lambda\pi(\omega)^{-1} d\omega \cdot \gamma^{-1} d\gamma.
\end{aligned} \tag{6.11}$$

Remember that this expression takes place in $\Omega^{del}_{\bullet}(\mathbb{C}\Gamma)_{ab}$, where

$$\begin{aligned}
\lambda\pi(\omega)^{-1} d\omega \cdot \gamma^{-1} d\gamma &= \pi(\omega)^{-1} d\omega \cdot \gamma^{-1} d\gamma \cdot \lambda \\
&= \pi(\omega)^{-1} d\omega \cdot (\gamma^{-1} d(\gamma\lambda) - d\lambda) \\
&= (-1)^{deg(d\omega)} (\gamma^{-1} d(\gamma\lambda) - d\lambda) \cdot \pi(\omega)^{-1} d\omega.
\end{aligned}$$

Therefore the third term of 6.11 is equal to

$$\begin{aligned}
&- \int_{x\in\mathcal{F}} \sum_{\lambda\neq e,\omega,\gamma} \mathrm{Tr}\left(R^*_{\gamma^{-1}} h(x) R^*_{(\lambda,e)} T_\omega(x,x)\right) \gamma^{-1} d(\gamma\lambda) \cdot \pi(\omega)^{-1} d\omega \\
&\qquad + \sum_{\lambda\neq e,\omega} \mathrm{Tr}\left(R^*_{(\lambda,e)} T_\omega(x,x)\right) d\lambda \cdot \pi(\omega)^{-1} d\omega.
\end{aligned}$$

In turn, after changing the first summation over γ to the summation over $\mu = \gamma\lambda$, this is equal to

$$
\begin{aligned}
(6.12)\qquad & -\int_{x\in\mathcal{F}} \sum_{\lambda\neq e,\omega,\mu} \operatorname{Tr}\left(R^*_{\lambda\mu^{-1}} h(x) R^*_{(\lambda,e)} T_\omega(x,x)\right) \lambda\mu^{-1} d\mu\cdot\pi(\omega)^{-1}d\omega \\
& + \sum_{\lambda\neq e,\omega} \operatorname{Tr}\left(R^*_{(\lambda,e)} T_\omega(x,x)\right) d\lambda\cdot\pi(\omega)^{-1}d\omega.
\end{aligned}
$$

Now, observing that the first term in (6.12) is equal to the opposite of the first term in (6.11), it follows that (6.11) becomes

$$
\begin{aligned}
\mathrm{TR}^{del}_0([\nabla^{\mathrm{Lott}},\mathbf{T}]) &= \int_{x\in\mathcal{F}} \sum_{\lambda\neq e,\omega} \operatorname{Tr}\left(R^*_{(\lambda,e)} T_\omega(x,x)\right) \lambda d\pi(\omega)^{-1}d\omega \\
&\quad + \sum_{\lambda\neq e,\omega} \operatorname{Tr}\left(R^*_{(\lambda,e)} T_\omega(x,x)\right) d\lambda\cdot\pi(\omega)^{-1}d\omega \\
&= \int_{x\in\mathcal{F}} \sum_{\omega,\lambda\neq e} \operatorname{Tr}\left(R^*_{(\lambda,e)} T_\omega(x,x)\right) d(\lambda\cdot\pi(\omega)^{-1}d\omega) \\
&= d\mathrm{TR}^{del}_0(\mathbf{T}).
\end{aligned}
\tag{6.13}
$$

Notice that, due to the fact that the support of the kernel of T_ω is Γ-compact, the sums over λ are finite. □

PROPOSITION 6.11. *Let* $\mathbf{T}\in\Psi^{-\infty}_{\widehat{\Omega}_\bullet(\mathcal{A}\Gamma)}(\widetilde{M},\widetilde{E})$, *then*

$$
d\mathrm{TR}(\mathbf{T}) = \mathrm{TR}([\nabla^{\mathrm{Lott}},\mathbf{T}]) \in \widehat{\Omega}_\bullet(\mathcal{A}\Gamma)_{ab}.
$$

PROOF. The proof is analogous to the one of Proposition 6.10. Observe that, due to Remark 6.5, in view of the previous algebraic manipulations, the fact that Fréchet algebras are involved is not a problem here. □

COROLLARY 6.12. *If* $\mathbf{T}\in\Psi^{0}_{\widehat{\Omega}_\bullet(\mathcal{A}\Gamma)}(\widetilde{M},\widetilde{E})$, *then*

$$
d\mathrm{TR}^{del}(\mathbf{T}) = \mathrm{TR}^{del}([\nabla^{\mathrm{Lott}},\mathbf{T}]) \in \widehat{\Omega}^{del}_\bullet(\mathcal{A}\Gamma)_{ab}.
$$

6.3. Mapping the analytic surgery sequence to non-commutative de Rham homology

In this section we shall use the tools developed in Section 4 to map the analytic surgery exact sequence to the non-commutative de Rham homology exact sequence (6.2). We shall use ideas of Connes, Gorokhovsky and Lott, see [**19**, **25**].

Notice that the pair $\left(\Psi^0_{\Omega_\bullet(\mathcal{A}\Gamma)}(\widetilde{M},\widetilde{E}),\overline{\nabla}^{\mathrm{Lott}}\right)$ is not a complex, because the square of Lott's connection $\overline{\nabla}^{\mathrm{Lott}}$ is not zero. Nevertheless, the triple

$$
\left(\Psi^0_{\Omega_\bullet(\mathcal{A}\Gamma)}(\widetilde{M},\widetilde{E}),\overline{\nabla}^{\mathrm{Lott}},\Theta\right)
$$

does verify the hypotheses of [**19**, Lemma 9, III.3], which produces in a canonical way the following complex.

DEFINITION 6.13. Let us define $(\overline{\Omega}_\bullet(\Psi^0_{\mathcal{A}\Gamma}(\widetilde{M},\widetilde{E})),d)$ as the following DGA over $\Psi^0_{\mathcal{A}\Gamma}(\widetilde{M},\widetilde{E})$:

- as a vector space, $\overline{\Omega}_\bullet(\Psi^0_{\mathcal{A}\Gamma}(\widetilde{M},\widetilde{E}))$ is

$$\Psi^0_{\widehat{\Omega}_\bullet(\mathcal{A}\Gamma)}(\widetilde{M},\widetilde{E}) \oplus X\cdot\Psi^0_{\widehat{\Omega}_\bullet(\mathcal{A}\Gamma)}(\widetilde{M},\widetilde{E}) \oplus \Psi^0_{\widehat{\Omega}_\bullet(\mathcal{A}\Gamma)}(\widetilde{M},\widetilde{E})\cdot X \oplus X\cdot\Psi^0_{\widehat{\Omega}_\bullet(\mathcal{A}\Gamma)}(\widetilde{M},\widetilde{E})\cdot X$$

  where X is a degree 1 auxiliary variable.
- The algebra structure is defined by setting $X^2=\Theta$ and $\mathbf{T}_1X\mathbf{T}_2=0$ for all $\mathbf{T}_i\in\Psi^0_{\widehat{\Omega}_\bullet(\mathcal{A}\Gamma)}(\widetilde{M},\widetilde{E})$.
- The differential d is defined by $dX=0$ and $d\mathbf{T}=\overline{\nabla}^{\mathrm{Lott}}\mathbf{T}+X\mathbf{T}+(-1)^{\deg\mathbf{T}}\mathbf{T}X$. One easily checks that $d^2=0$.

The same construction applies to $\left(\Psi^{-\infty}_{\widehat{\Omega}_\bullet(\mathcal{A}\Gamma)}(\widetilde{M},\widetilde{E}),\overline{\nabla}^{\mathrm{Lott}}\right)$ and we obtain a subcomplex $\overline{\Omega}_\bullet(\Psi^{-\infty}_{\mathcal{A}\Gamma}(\widetilde{M},\widetilde{E}))$ of $\overline{\Omega}_\bullet(\Psi^0_{\mathcal{A}\Gamma}(\widetilde{M},\widetilde{E}))$.

Now we want to extend form valued traces to these new complexes.

DEFINITION 6.14. Let us consider an element of $\overline{\Omega}_\bullet(\Psi^0_{\mathcal{A}\Gamma}(\widetilde{M},\widetilde{E}))$, it is of the form $\mathbf{T}_{11}+\mathbf{T}_{12}X+X\mathbf{T}_{21}+X\mathbf{T}_{22}X$. Then set

$$\overline{\mathrm{TR}}^{del}(\mathbf{T}_{11}+\mathbf{T}_{12}X+X\mathbf{T}_{21}+X\mathbf{T}_{22}X) := \mathrm{TR}^{del}(\mathbf{T}_{11})-(-1)^{\deg\mathbf{T}_{22}}\mathrm{TR}^{del}(\mathbf{T}_{22}\Theta)\in\widehat{\Omega}^{del}_\bullet(\mathcal{A}\Gamma)_{ab}. \tag{6.14}$$

The extension $\overline{\mathrm{TR}}\colon\overline{\Omega}_\bullet(\Psi^{-\infty}_{\mathcal{A}\Gamma}(\widetilde{M},\widetilde{E}))\to\widehat{\Omega}_\bullet(\mathcal{A}\Gamma)_{ab}$ of TR is defined analogously.

LEMMA 6.15. *The following map, defined as in* (6.14),

$$\overline{\mathrm{TR}}^{del}\colon\overline{\Omega}_\bullet(\Psi^0_{\mathcal{A}\Gamma}(\widetilde{M},\widetilde{E}))_{ab}\to\widehat{\Omega}^{del}_\bullet(\mathcal{A}\Gamma)_{ab}$$

is a morphism of complexes. The same is true for $\overline{\mathrm{TR}}\colon\overline{\Omega}_\bullet(\Psi^{-\infty}_{\mathcal{A}\Gamma}(\widetilde{M},\widetilde{E}))_{ab}\to\widehat{\Omega}_\bullet(\mathcal{A}\Gamma)_{ab}$.

PROOF. Let $\mathbf{T}_{11}+\mathbf{T}_{12}X+X\mathbf{T}_{21}+X\mathbf{T}_{22}X$ be an element of $\overline{\Omega}_\bullet(\Psi^0_{\mathcal{A}\Gamma}(\widetilde{M},\widetilde{E}))$, then by Proposition 6.10 we have

$$\begin{aligned}
&\overline{\mathrm{TR}}^{del}(d(\mathbf{T}_{11}+\mathbf{T}_{12}X+X\mathbf{T}_{21}+X\mathbf{T}_{22}X))\\
&=\mathrm{TR}^{del}\left([\nabla^{\mathrm{Lott}},\mathbf{T}_{11}]\right)-(-1)^{\deg\mathbf{T}_{22}}\mathrm{TR}^{del}\left([\nabla^{\mathrm{Lott}},\mathbf{T}_{22}]\Theta\right)\\
&=\mathrm{TR}^{del}\left([\nabla^{\mathrm{Lott}},\mathbf{T}_{11}]\right)-(-1)^{\deg\mathbf{T}_{22}}\mathrm{TR}^{del}\left([\nabla^{\mathrm{Lott}},\mathbf{T}_{22}\Theta]\right)\\
&=d\mathrm{TR}^{del}(\mathbf{T}_{11})-(-1)^{\deg\mathbf{T}_{22}}d\mathrm{TR}^{del}(\mathbf{T}_{22}\Theta)\\
&=d\overline{\mathrm{TR}}^{del}(\mathbf{T}_{11}+\mathbf{T}_{12}X+X\mathbf{T}_{21}+X\mathbf{T}_{22}X).
\end{aligned}\tag{6.15}$$

and hence the lemma follows. The proof for $\overline{\mathrm{TR}}$ is identical. □

So far, we have used the algebras of smoothing operators $\Psi^{-\infty}_{\mathbb{C}\Gamma}(\widetilde{M},\widetilde{E})$ and of pseudodifferential operators of order 0, $\Psi^0_{\mathbb{C}\Gamma}(\widetilde{M},\widetilde{E})$, and versions with coefficients in non-commutative differential forms over $\mathbb{C}\Gamma$, as well as Fréchet completions $\mathcal{A}\Gamma$, and also the X-extended DGA $\overline{\Omega}_\bullet(\Psi^0_{\mathcal{A}\Gamma}(\widetilde{M},\widetilde{E}))$.

Observe that we can of course also pass to the versions of pseudodifferential operators of arbitrary, possibly positive, degree. These are not contained in bounded operators on L^2-sections (and the associated C^*-algebra), but the whole constructions make sense in the context of abstract algebras and even Fréchet algebras.

PROPOSITION 6.16. *The connection* $\overline{\nabla}^{\mathrm{Lott}}$ *and the delocalized trace* TR^{del} *and all the other constructions and results of Section 6 derived so far extend to the versions of pseudodifferential operators* $\Psi^{+\infty}$ *of arbitrary order. In particular,* $\mathrm{TR}^{del}\colon \Psi^{+\infty}_{\hat{\Omega}_\bullet(\mathcal{A}\Gamma)}(\widetilde{M},\widetilde{E}) \to \widehat{\Omega}^{del}_\bullet(\mathcal{A}\Gamma)_{ab}$ *has the trace property.*

PROOF. This follows with exactly the same arguments as in the case of operators of order 0. The crucial point is that pseudodifferential operators of arbitrary order have Schwarz kernels which are smooth outside the diagonal. The singularities on the diagonal are "worse" for operators of positive order than for operators of order 0, but the diagonal anyway is cut off in the relevant constructions □

LEMMA 6.17. *Let* $\overline{\Omega}_\bullet(C^\infty(M),E)$ *be the subcomplex of* $\overline{\Omega}_\bullet(\Psi^0_{\mathcal{A}\Gamma}(\widetilde{M},\widetilde{E}))$ *generated by* $C^\infty(M,E)$ *(acting as multiplication operators), then we have that*

(1) $\overline{\Omega}_*(C^\infty(M,E))\cap\overline{\Omega}_*(\Psi^{-\infty}_{\mathcal{A}\Gamma}(\widetilde{M},\widetilde{E}))=\{0\}$,

(2) $\overline{\mathrm{TR}}^{del}$ *sends* $\overline{\Omega}_*(C^\infty(M,E))$ *to zero.*

PROOF. First observe that $\overline{\nabla}^{\mathrm{Lott}}f$ is zero for any smooth function f on M. This implies that the subcomplex generated by $C^\infty(M)$ consists of linear combinations of products of elements of $C^\infty(M)$ and X. Therefore, the support of any element in $\overline{\Omega}_*(C^\infty(M,E))$ is contained in the diagonal of $\widetilde{M}\times\widetilde{M}$. On the other hand, a smoothing operator with support on the diagonal is zero. Hence, the assertions follow. □

LEMMA 6.18. *The DGA* $\overline{\Omega}_\bullet(\Psi^{-\infty}_{\mathcal{A}\Gamma}(\widetilde{M},\widetilde{E}))$ *is Fréchet split excisive.*

PROOF. First let us consider the algebra of smoothing operators $\Psi^{-\infty}(M)$. Taking the Schwartz kernel gives an isomorphism with $C^\infty(M\times M)$, where the composition is given by convolution. A way of seeing this convolution is the following one. Consider:

- the smooth inclusion $i\colon M\times M\times M\to(M\times M)\times(M\times M)$ given by $(x,z,y)\mapsto(x,z,z,y)$;
- the smooth projection $p\colon M\times M\times M\to M\times M$ given by $(x,z,y)\mapsto(x,y)$.

After fixing a Riemannian measure μ on M, the convolution of two smoothing kernels k_1 and k_2 is given by the following formula

$$k_1 * k_2 = p_*(i^*(k_1\otimes k_2))$$

where i^* is the restriction induced by the pull-back and p_* is the integration along the fibers of the surjection.

Now, let us fix $k\in C^\infty(M\times M)$. If we put $\overline{k}(x,z,y):=(\mathrm{Vol}(M))^{-1}k(x,y)$, then $p_*(\overline{k})=k$. Moreover, since i is injective, $i^*\colon C^\infty(M\times M\times M\times M)\to C^\infty(M\times M\times M)$ is surjective. By using the fact that $C^\infty(M\times M\times M\times M)\cong C^\infty(M\times M)\hat{\otimes}C^\infty(M\times M)$ we find then $k_i^1,k_i^2\in C^\infty(M\times M)$ for $i\in\mathbb{N}$ such that $\overline{k}=\sum_i k_i^1\otimes k_i^2$. Hence we finally have that

$$k=\sum_{i\in\mathbb{N}}k_i^1 * k_i^2\in C^\infty(M\times M)\quad\forall k\in C^\infty(M\times M).$$

Now we consider the algebra $\Psi^\infty_{\hat{\Omega}_\bullet(\mathcal{A}\Gamma)}(\widetilde{M},\widetilde{E})$ of smoothing operators with coefficients in forms. As explained in Lemma 5.3 these are given by smooth functions on $C^\infty(M\times M,pr_1^*F\otimes pr_2^*F^*)$, for a certain bundle of modules F over M. In this case the argument is similar to the previous one, just by using the fact that the composition $(pr_1^*F \otimes pr_2^*F^*) \otimes (pr_1^*F \otimes pr_2^*F^*) \to pr_1^*F \otimes pr_2^*F^*$ is surjective. Finally, by using this argument on each summand of $\overline{\Omega}_\bullet(\Psi^{-\infty}_{\mathcal{A}\Gamma}(\widetilde{M},\widetilde{E}))$, we obtain the desired result. $\square$

We are now in the position to define our maps from the analytic surgery exact sequence in Subsection 2.1,

$$\begin{aligned}(6.16)\qquad \ldots \xrightarrow{\partial} K_*(0 \hookrightarrow C^*_{red}(\widetilde{M}\times_\Gamma \widetilde{M})) \xrightarrow{i_*} K_*(C(M) \xrightarrow{\mathrm{m}} \Psi^0_\Gamma(\widetilde{M}))\\ \xrightarrow{\sigma_*} K_*(C(M) \xrightarrow{\pi^*} C(S^*M)) \xrightarrow{\partial} \cdots\end{aligned}$$

to

(6.17)

$$\cdots \longrightarrow H_*(\mathcal{A}\Gamma) \longrightarrow H^{del}_*(\mathcal{A}\Gamma) \xrightarrow{\delta_\Gamma} H^e_{*+1}(\mathcal{A}\Gamma) \longrightarrow H_{*+1}(\mathcal{A}\Gamma) \longrightarrow \cdots$$

This will employ Definitions 4.13 and 4.19 for the relative Chern characters. Notice that we have expunged the vector bundles from the notation, given that from the K-theoretic point of view they are negligible.

We first observe that there is a canonical isomorphism of long exact sequences

$$\begin{array}{ccccc}
\cdots K_*(0 \to \Psi^{-\infty}_{\mathcal{A}\Gamma}(\widetilde{M})) & \xrightarrow{i_*} & K_*(C^\infty(M) \xrightarrow{\mathrm{m}} \Psi^0_{\mathcal{A}\Gamma}(\widetilde{M})) & \xrightarrow{\sigma_*} & K_*(C^\infty(M) \to \Psi^0_{\mathcal{A}\Gamma}(\widetilde{M})/\Psi^{-\infty}_{\mathcal{A}\Gamma}(\widetilde{M})) \\
\downarrow & & \downarrow & & \downarrow \\
K_*(0 \to C^*_{red}(\widetilde{M}\times_\Gamma \widetilde{M})) & \xrightarrow{i_*} & K_*(C(M) \xrightarrow{\mathrm{m}} \Psi^0_\Gamma(\widetilde{M})) & \xrightarrow{\sigma_*} & K_*(C(M) \xrightarrow{\pi^*} C(S^*M)) \cdots
\end{array}$$

The first two vertical arrows are isomorphism because they are induced by inclusions of dense holomorphically closed subalgebras. The third homomorphism is induced by the symbol map and it is an isomorphism because of the Five Lemma.

We can think of the exact sequence

$$\begin{aligned}\cdots \xrightarrow{\partial} K_*(0 \to \Psi^{-\infty}_{\mathcal{A}\Gamma}(\widetilde{M})) \xrightarrow{i_*} K_*(C^\infty(M) \xrightarrow{\mathrm{m}} \Psi^0_{\mathcal{A}\Gamma}(\widetilde{M}))\\ \xrightarrow{\sigma_*} K_*(C^\infty(M) \to \Psi^0_{\mathcal{A}\Gamma}(\widetilde{M})/\Psi^{-\infty}_{\mathcal{A}\Gamma}(\widetilde{M})) \xrightarrow{\partial} \cdots\end{aligned}$$

as a smooth version of the analytic surgery sequence. It is this smooth version that will be mapped to non-commutative de Rham homology.

DEFINITION 6.19.

(1) Let x be a class in $K_*(0 \to \Psi^{-\infty}_{\mathcal{A}\Gamma}(\widetilde{M}))$. Then define

$$\mathrm{Ch}_\Gamma(x) := \overline{\mathrm{TR}}_{[*-1]}\left(\mathrm{Ch}^{rel}(x)\right) \in H_{[*-1]}(\mathcal{A}\Gamma).$$

Here

$$(6.18)\qquad \mathrm{Ch}^{rel}\colon K_*(0 \to \Psi^{-\infty}_{\mathcal{A}\Gamma}(\widetilde{M})) \to H_{[*-1]}\left(0 \to \overline{\Omega}(\Psi^{-\infty}_{\mathcal{A}\Gamma}(\widetilde{M}))\right),$$

and $\overline{\mathrm{TR}}_*$ is a compact notation for

$$(0,\overline{\mathrm{TR}})_*\colon H_*\left(0 \to \overline{\Omega}(\Psi^{-\infty}_{\mathcal{A}\Gamma}(\widetilde{M}))\right) \to H_*(\mathcal{A}\Gamma).$$

(2) Let y be a class in $K_*(C^\infty(M) \to \Psi^0_{\mathcal{A}\Gamma}(\widetilde{M}))$. Then define

$$\mathrm{Ch}^{del}_\Gamma(y) := \overline{\mathrm{TR}}^{del}_{[*-1]}\left(\mathrm{Ch}^{rel}(y)\right) \in H^{del}_{[*-1]}(\mathcal{A}\Gamma).$$

Here

$$\mathrm{Ch}^{rel}\colon K_*(C^\infty(M) \to \Psi^0_{\mathcal{A}\Gamma}(\widetilde{M})) \to H_{[*-1]}\left(\overline{\Omega}(C^\infty(M)) \to \overline{\Omega}(\Psi^0_{\mathcal{A}\Gamma}(\widetilde{M}))\right)$$

and $\overline{\mathrm{TR}}^{del}_*$ is a compact notation for

$$(0, \overline{\mathrm{TR}}^{del})_*\colon H_*\left(\overline{\Omega}(C^\infty(M)) \to \overline{\Omega}(\Psi^0_{\mathcal{A}\Gamma}(\widetilde{M}))\right) \to H^{del}_*(\mathcal{A}\Gamma).$$

(3) Let z be a class in $K_*(C^\infty(M) \to \Psi^0_{\mathcal{A}\Gamma}(\widetilde{M})/\Psi^{-\infty}_{\mathcal{A}\Gamma}(\widetilde{M}))$. Recall from Lemma 4.27 that

$$q_*\colon K_*(C^\infty(M) + \Psi^{-\infty}_{\mathcal{A}\Gamma}(\widetilde{M}) \to \Psi^0_{\mathcal{A}\Gamma}(\widetilde{M})) \to K_*(C^\infty(M) \to \Psi^0_{\mathcal{A}\Gamma}(\widetilde{M})/\Psi^{-\infty}_{\mathcal{A}\Gamma}(\widetilde{M}))$$

is an isomorphism. Then define

$$\mathrm{Ch}^e_\Gamma(z) := J_*^{-1} \circ (\overline{\mathrm{TR}}, \overline{\mathrm{TR}}^{del})_{[*-1]}\left(\mathrm{Ch}^{rel}(q_*^{-1}(z))\right) \in H^e_{[*]}(\mathcal{A}\Gamma).$$

Here

$$\begin{aligned}\mathrm{Ch}^{rel}\colon K_*(C^\infty(M) + \Psi^{-\infty}_{\mathcal{A}\Gamma}(\widetilde{M}) &\to \Psi^0_{\mathcal{A}\Gamma}(\widetilde{M})) \\ &\to H_*\left(\overline{\Omega}(C^\infty(M)) + \overline{\Omega}(\Psi^{-\infty}_{\mathcal{A}\Gamma}(\widetilde{M})) \to \overline{\Omega}(\Psi^0_{\mathcal{A}\Gamma}(\widetilde{M}))\right)\end{aligned}$$

and

$$\begin{aligned}(\overline{\mathrm{TR}}, \overline{\mathrm{TR}}^{del})\colon &\left(\overline{\Omega}(C^\infty(M)) + \overline{\Omega}(\Psi^{-\infty}_{\mathcal{A}\Gamma}(\widetilde{M})) \to \overline{\Omega}(\Psi^0_{\mathcal{A}\Gamma}(\widetilde{M}))\right) \\ &\to \left(\widehat{\Omega}(\mathcal{A}\Gamma) \to \widehat{\Omega}^{del}(\mathcal{A}\Gamma)\right)\end{aligned}$$

is the natural morphism of mapping cone complexes induced by $\overline{\mathrm{TR}}$ and $\overline{\mathrm{TR}}^{del}$; the isomorphism J_* (which is of degree -1) is induced by the inclusion J of $\widehat{\Omega}^e(\mathcal{A}\Gamma)$ into $\left(\widehat{\Omega}(\mathcal{A}\Gamma) \to \widehat{\Omega}^{del}(\mathcal{A}\Gamma)\right)$. See the proof of Theorem 6.20 below for more on J_*.

Observe that we can extend the map of complexes $\overline{\mathrm{TR}}$ to $\overline{\Omega}(C^\infty) + \overline{\Omega}(\Psi^{-\infty}_{\mathcal{A}\Gamma})$ by setting it to be zero on $\overline{\Omega}(C^\infty)$. Indeed, although it is not a direct sum of DGAs, $\overline{\Omega}(C^\infty) + \overline{\Omega}(\Psi^{-\infty}_{\mathcal{A}\Gamma})$ is a direct sum of complexes and then this extension makes sense, using Lemma 6.15.
The following theorem is one of the main results of this article.

THEOREM 6.20. *The following diagram, with vertical maps as in Definition* 6.19, *is commutative:*
(6.19)

$$\begin{array}{ccccc}
\to K_*(0 \to \Psi^{-\infty}_{\mathcal{A}\Gamma}(\widetilde{M})) & \xrightarrow{i_*} & K_*(C^\infty(M) \xrightarrow{\mathrm{m}} \Psi^0_{\mathcal{A}\Gamma}(\widetilde{M})) & \xrightarrow{\sigma_*} & K_*(C^\infty(M) \xrightarrow{\pi^*} \Psi^0_{\mathcal{A}\Gamma}(\widetilde{M})/\Psi^{-\infty}_{\mathcal{A}\Gamma}(\widetilde{M})) \to \\
\downarrow{\scriptstyle \mathrm{Ch}_\Gamma} & & \downarrow{\scriptstyle \mathrm{Ch}^{del}_\Gamma} & & \downarrow{\scriptstyle \mathrm{Ch}^e_\Gamma} \\
\to H_{[*\ 1]}(\mathcal{A}\Gamma) & \longrightarrow & H^{del}_{[*\ 1]}(\mathcal{A}\Gamma) & \xrightarrow{\delta} & H^e_{[*]}(\mathcal{A}\Gamma) \to
\end{array}$$

PROOF. In the following, we write C^∞, $\Psi^0_{\mathcal{A}\Gamma}$ and $\Psi^{-\infty}_{\mathcal{A}\Gamma}$ as shorthands for $C^\infty(M)$, $\Psi^0_{\mathcal{A}\Gamma}(\widetilde{M})$ and $\Psi^{-\infty}_{\mathcal{A}\Gamma}(\widetilde{M})$, respectively. By Lemma 4.27, we have the following isomorphism of long exact sequences
(6.20)

$$\begin{CD}
\cdots @>{\partial'}>> K_{*+1}(\Psi^{-\infty}_{\mathcal{A}\Gamma}) @>{S}>> K_*(C^\infty \to \Psi^0_{\mathcal{A}\Gamma}) @>>> K_*(C^\infty + \Psi^{-\infty}_{\mathcal{A}\Gamma} \to \Psi^0_{\mathcal{A}\Gamma}) @>{\partial'}>> \cdots \\
@. @VV{S}V @VV{\mathrm{id}}V @VV{q_*}V @. \\
\cdots @>{\partial}>> K_*(0 \to \Psi^{-\infty}_{\mathcal{A}\Gamma}) @>>> K_*(C^\infty \to \Psi^0_{\mathcal{A}\Gamma}) @>>> K_*(C^\infty \to \Psi^0_{\mathcal{A}\Gamma}(\widetilde{M})/\Psi^{-\infty}_{\mathcal{A}\Gamma}(\widetilde{M})) @>{\partial}>> \cdots
\end{CD}$$

and by Lemma 6.18 by Theorem 4.32 there exists the following commutative diagram
(6.21)

$$\begin{CD}
\to K_{*+1}(\Psi^{-\infty}_{\mathcal{A}\Gamma}) @>{S}>> K_*(C^\infty \to \Psi^0_{\mathcal{A}\Gamma}) @>>> K_*(C^\infty + \Psi^{-\infty}_{\mathcal{A}\Gamma} \to \Psi^0_{\mathcal{A}\Gamma}) \to \\
@VV{\mathrm{Ch}}V @VV{\mathrm{Ch}^{rel}}V @VV{\mathrm{Ch}^{rel}}V \\
H_{[*+1]}(\overline{\Omega}(\Psi^{-\infty}_{\mathcal{A}\Gamma})) @>{S}>> H_{[*+1]}(\overline{\Omega}(C^\infty) \to \overline{\Omega}(\Psi^0_{\mathcal{A}\Gamma})) @>>> H_{[*+1]}(\overline{\Omega}(C^\infty) + \overline{\Omega}(\Psi^{-\infty}_{\mathcal{A}\Gamma}) \to \overline{\Omega}(\Psi^0_{\mathcal{A}\Gamma})) \to
\end{CD}$$

Furthermore the following commutative diagram of complexes
(6.22)

$$\begin{CD}
0 \to (\overline{\Omega}_*(C^\infty) \to \overline{\Omega}_*(\Psi^0_{\mathcal{A}\Gamma})) @>>> (\overline{\Omega}_*(C^\infty) + \overline{\Omega}_*(\Psi^{-\infty}_{\mathcal{A}\Gamma}) \to \overline{\Omega}_*(\Psi^0_{\mathcal{A}\Gamma})) @>>> \overline{\Omega}_{*+1}(\Psi^{-\infty}_{\mathcal{A}\Gamma}) \to 0 \\
@VV{\mathrm{TR}^{del}}V @VV{(\mathrm{TR},\mathrm{TR}^{del})}V @VV{\mathrm{TR}}V \\
0 \to \widehat{\Omega}^{del}_*(\mathcal{A}\Gamma) @>>> (\widehat{\Omega}_*(\mathcal{A}\Gamma) \to \widehat{\Omega}^{del}_*(\mathcal{A}\Gamma)) @>>> \widehat{\Omega}_{*+1}(\mathcal{A}\Gamma) \to 0
\end{CD}$$

induces a map of long exact sequences in homology. Finally observe that we have a further isomorphism of exact sequences given by
(6.23)

$$\begin{CD}
\cdots @>>> H_*(\widehat{\Omega}^{del}(\mathcal{A}\Gamma)) @>{\delta_\Gamma}>> H_{*+1}(\widehat{\Omega}^e(\mathcal{A}\Gamma)) @>>> H_{*+1}(\widehat{\Omega}(\mathcal{A}\Gamma)) @>>> \cdots \\
@. @VV{=}V @V{\cong}V{J_*}V @VV{=}V @. \\
\cdots @>>> H_*(\widehat{\Omega}^{del}(\mathcal{A}\Gamma)) @>>> H_*(\widehat{\Omega}(\mathcal{A}\Gamma) \to \widehat{\Omega}^{del}(\mathcal{A}\Gamma)) @>>> H_{*+1}(\widehat{\Omega}(\mathcal{A}\Gamma)) @>>> \cdots
\end{CD}$$

where the arrow in the middle is induced by the inclusion of $\widehat{\Omega}^e_*(\mathcal{A}\Gamma)$ into $(\widehat{\Omega}_*(\mathcal{A}\Gamma) \to \widehat{\Omega}^{del}_*(\mathcal{A}\Gamma))$. As the mapping of exact sequences in (6.19) is given by the composition of the inverse of (6.20) followed by (6.21), (6.22) and the inverse of (6.23), the result follows. $\square$

By using the isomorphism of long exact sequence in (2.2) and passing from C*-algebras to smooth dense subalgebra, thanks to the previous theorem, we obtain the following commutative diagram

$$\begin{CD}
\cdots @>>> K_{*-1}(C^*_{red}\Gamma) @>{s}>> \mathrm{S}^\Gamma_*(\widetilde{M}) @>{c}>> K^\Gamma_*(\widetilde{M}) @>>> \cdots \\
@. @VV{\mathrm{Ch}_\Gamma}V @VV{\mathrm{Ch}^{del}_\Gamma}V @VV{\mathrm{Ch}^e_\Gamma}V @. \\
\cdots @>{j_*}>> H_{[*-1]}(\mathcal{A}\Gamma) @>>> H^{del}_{[*-1]}(\mathcal{A}\Gamma) @>{\delta}>> H^e_{[*]}(\mathcal{A}\Gamma) @>{j_*}>> \cdots
\end{CD} \tag{6.24}$$

which is the result stated in the Introduction.

6.4. The algebraic index and the localized Chern character Ch^e_Γ

In the next section we will deal with the delocalized Chern character Ch^{del}_Γ. In the current subsection, we want to recognize the localized Ch^e_Γ as a well known homomorphism involving the algebraic index.

REMARK 6.21. Before proceeding with this section, we need to recall some facts about relative algebraic K-theory, without entering into the details. Observe that the equivalence relation used in the algebraic world between two idempotent e_0 and e_1 over A is to be stably conjugated by an invertible element z over A. Hence the elements in the relative K-group $K_0^{alg}(A \xrightarrow{\varphi} B)$ are given by triples $[e_0, e_1; z]$, where e_0, e_1 are idempotents over A and z is an invertible element over B which conjugate $\varphi(e_0)$ into $\varphi(e_1)$. If A is a dense holomorphically closed subalgebra of a C^*-algebra, this definition of relative K-theory is equivalent to the one used so far.

In the algebraic setting, given two idempotent e_0, e_1 which are conjugated by an invertible element z, they are homotopic through a path f_t defined as in (3.1). This is sufficient to define

$$\mathrm{Ch}^{rel}_{alg}\colon K_0^{alg}(A \to B) \to H_{even}(\Omega(A) \to \Omega(B)) \tag{6.25}$$

exactly as in Definition 4.13: notice indeed that $\int_0^1 \overline{\mathrm{Ch}}(f_t)dt$ can be explicitly calculated and it is given by a completely algebraic expression in e_0, e_1 and z.

We now recall the definition of the algebraic index, see for instance [**20**, Definition 5.2]. First we remark that

$$\Sigma := \Psi^0_{\mathcal{A}\Gamma}(\widetilde{M})/\Psi^{-\infty}_{\mathcal{A}\Gamma}(\widetilde{M}) \cong \Psi^0_{\mathbb{C}\Gamma}(\widetilde{M})/\Psi^{-\infty}_{\mathbb{C}\Gamma}(\widetilde{M}),$$

the algebra of complete symbols. The isomorphism is canonical and it is induced by the inclusion $\Psi^0_{\mathbb{C}\Gamma} \to \Psi^0_{\mathcal{A}\Gamma}$. Let us see the details. The map is injective, because $\Psi^0_{\mathbb{C}\Gamma}(\widetilde{M}) \cap \Psi^{-\infty}_{\mathcal{A}\Gamma}(\widetilde{M}) = \Psi^{-\infty}_{\mathbb{C}\Gamma}(\widetilde{M})$, the operators which are at the same time smoothing and satisfy the finite support condition of $\Psi_{\mathbb{C}\Gamma}(\widetilde{M})$. For the surjectivity, we use the cutoff argument already employed in Definition 6.8 of the delocalized trace. Indeed, there we saw that every $\mathbf{T} \in \Psi^0_{\mathcal{A}\Gamma}(\widetilde{M})$ can be written as a sum $\mathbf{T} = \mathbf{T}_0 + \mathbf{T}_\infty$ with $\mathbf{T}_0 \in \Psi^0_{\mathbb{C}\Gamma}$ and $\mathbf{T}_\infty \in \Psi^{-\infty}_{\mathcal{A}\Gamma}$, showing that the inclusion induces a surjective map $\Psi^0_{\mathbb{C}\Gamma}(\widetilde{M}) \to \Psi^0_{\mathcal{A}\Gamma}(\widetilde{M})/\Psi^{-\infty}_{\mathcal{A}\Gamma}(\tilde{M})$, as required.

The algebraic index

$$\mathrm{Ind}_{\mathbb{C}\Gamma}\colon K_0(C^\infty(M) \to \Sigma) \to K_0(\Psi^{-\infty}_{\mathbb{C}\Gamma}(\widetilde{M})) \tag{6.26}$$

is defined in the following way. A class in $K_0(C^\infty(M) \to \Sigma)$ is given by a triple $[e_0, e_1; \sigma]$, where $e_i \in M_n(C^\infty(M))$ is the projection associated to the vector bundle E_i, and where σ is an invertible complete symbol such that $e_0 = \sigma^{-1}e_1\sigma$, here we identify e_i with its image in the complete symbol algebra. Choose two pseudodifferential operators $P \in \Psi^0_{\mathbb{C}\Gamma}(\widetilde{M}; \widetilde{E}_0, \widetilde{E}_1)$ and $Q \in \Psi^0_{\mathbb{C}\Gamma}(\widetilde{M}; \widetilde{E}_1, \widetilde{E}_0)$ corresponding to σ or σ^{-1}, respectively. Concretely, $P = e_1 A e_0$ and $Q = e_0 B e_1$ where $A, B \in \Psi^0_{\mathbb{C}\Gamma}(\widetilde{M}, \mathbb{C}^n)$ with symbols σ, σ^{-1}. Then

$$S = 1_{E_0} - QP, \quad T = 1_{E_1} - PQ \in \Psi^{-\infty}_{\mathbb{C}\Gamma}(\widetilde{M}, \widetilde{E}_0 \oplus \widetilde{E}_1)$$

and

$$\mathrm{Ind}_{\mathbb{C}\Gamma}([e_0, e_1; \sigma]) := [CM(P,Q)] - [diag(0, 1_{E_1})] \in K_0(\Psi^{-\infty}_{\mathbb{C}\Gamma}(\widetilde{M})) \tag{6.27}$$

where

$$CM(P,Q) := \begin{pmatrix} S^2 & S(1_{E_0}+S)Q \\ TP & 1_{E_1}-T^2 \end{pmatrix} \tag{6.28}$$

as in [**20**, Section 2].

Let Ch denote the first vertical map in (6.21).

LEMMA 6.22. *The image of* $\overline{\mathrm{TR}}_* \circ \mathrm{Ch} \circ \mathrm{Ind}_{\mathbb{C}\Gamma}$ *belongs to*

$$H_{\mathrm{even}}(\Omega^e_*(\mathbb{C}\Gamma)) \subset H_{\mathrm{even}}(\Omega_*(\mathbb{C}\Gamma)).$$

PROOF. This follows immediately from the fact that we can make our choices so that $CM(P,Q)$ is almost local and that the class does not depend on choices. □

Using Lemma 6.22, we can give the following definition.

DEFINITION 6.23.

$$\mathrm{Ch}^e_{\Gamma,alg} := \overline{\mathrm{TR}}_* \circ \mathrm{Ch} \circ \mathrm{Ind}_{\mathbb{C}\Gamma} \colon K_0(C^\infty(M) \to \Sigma) \to H^e_{\mathrm{even}}(\mathbb{C}\Gamma). \tag{6.29}$$

In fact, we can say much more. Define a Dirac class $[e_0, e_1; \sigma]$ as the class associated to a generalized Dirac operator.

PROPOSITION 6.24. *If* $[e_0, e_1; \sigma] \in K_0(C^\infty(M) \to \Sigma)$ *is a Dirac class, then*

$$\mathrm{Ch}^e_{\Gamma,alg}([e_0, e_1; \sigma]) = \left[\int_M \mathrm{AS}(M,\sigma) \wedge \omega_{\mathrm{Lott}}\right] \in H^e_{\mathrm{even}}(\mathbb{C}\Gamma),$$

where $\mathrm{AS}(M,\sigma)$ *is the corresponding Atiyah-Singer integrand and where* $\omega_{\mathrm{Lott}}(M) \in \Omega^*(M) \otimes \Omega_*(\mathbb{C}\Gamma)$ *is Lott's bi-form.*

For more on $\omega_{\mathrm{Lott}}(M)$, a bi-form which is closed in both arguments and that appears naturally in Lott's treatment of the Connes-Moscovici higher index theorem, we refer the reader to [**51**] and also [**44**, Theorem 13.6].

PROOF. Taking into account the characterization of $\mathrm{Ch}^e_{\Gamma,alg}$ given in the previous Lemma, the result follows from a sharpening of Lott's higher index theorem for operators of Dirac type. Indeed, for this particular statement we need to employ the version of Lott's theorem given in [**25**, Section 5] (just take B equal to a point there), but with the improvement given in [**26**, Appendix A], which allows to prove this formula by using almost local parametrices and the Getzler calculus. Notice that this argument avoids heat kernel techniques and rests ultimately on the analysis given by Moscovici and Wu in [**62**]. In these references the left hand side is paired with a cyclic cocycle; however, as explained in [**48**], see in particular Definition 11 and Theorem 4 (1) there, one can give a statement directly at the level of non-commutative de Rham classes. □

It is straightforward to see that the formula (6.27) for $\mathrm{Ind}_{\mathbb{C}\Gamma}$ is an explicit implementation of the composition of $K_0(C^\infty(M) \to \Sigma) \to K_0(\Psi^0_{\mathbb{C}\Gamma}(\widetilde{M}) \to \Sigma)$ and the excision isomorphism $K_0(\Psi^{-\infty}_{\mathbb{C}\Gamma}(\widetilde{M})) \xrightarrow{\cong} K_0(\Psi^0_{\mathbb{C}\Gamma}(\widetilde{M}) \to \Sigma)$. Equivalently, $\mathrm{Ind}_{\mathbb{C}\Gamma}$ is the boundary map for the extension of pairs of algebras

$$0 \to (0 \to \Psi^{-\infty}_{\mathbb{C}\Gamma}(\widetilde{M})) \to (C^\infty(M) \to \Psi^0_{\mathbb{C}\Gamma}(\widetilde{M})) \to (C^\infty(M) \to \Sigma) \to 0,$$

composed with the suspension isomorphism. It is then a direct consequence of Theorem 4.32 that the following square is commutative, where we use that the top row is, by Lemma 4.27, the map ∂' of Theorem 4.32:
(6.30)

$$\begin{array}{ccccc} K_0(C^\infty + \Psi^{-\infty}_{\mathbb{C}\Gamma} \to \Psi^0_{\mathbb{C}\Gamma}) & \xrightarrow{\cong} & K_0(C^\infty \to \Sigma) & \xrightarrow{\mathrm{Ind}_{\mathbb{C}\Gamma}} & K_0(\Psi^{-\infty}_{\mathbb{C}\Gamma}) \\ \downarrow{\scriptstyle \mathrm{Ch}^{rel}} & & & & \downarrow{\scriptstyle \mathrm{Ch}} \\ H_{\mathrm{odd}}(\overline{\Omega}_*(C^\infty) + \overline{\Omega}_*(\Psi^{-\infty}_{\mathbb{C}\Gamma}) \to \overline{\Omega}_*(\Psi^0_{\mathbb{C}\Gamma})) & & \xrightarrow{\delta'_c = S \circ pr} & & H_{\mathrm{even}}(\overline{\Omega}_*(\Psi^{-\infty}_{\mathbb{C}\Gamma})). \end{array}$$

REMARK 6.25. Before stating the main proposition of this section, it is worth to notice that, by the very definition of the relative algebraic Chern character, see (6.25), the following square is commutative:

$$\begin{array}{ccc} K_0^{alg}(C^\infty + \Psi^{-\infty}_{\mathbb{C}\Gamma} \to \Psi^0_{\mathbb{C}\Gamma}) & \xrightarrow{\mathrm{Ch}^{rel}_{alg}} & H_{\mathrm{odd}}(\overline{\Omega}(C^\infty) + \overline{\Omega}(\Psi^{-\infty}_{\mathbb{C}\Gamma}) \to \overline{\Omega}(\Psi^0_{\mathbb{C}\Gamma})) \\ {\scriptstyle \cong}\downarrow{\scriptstyle i_*} & & \downarrow \\ K_0(C^\infty + \Psi^{-\infty}_{\mathcal{A}\Gamma} \to \Psi^0_{\mathcal{A}\Gamma}) & \xrightarrow{\mathrm{Ch}^{rel}} & H_{\mathrm{odd}}(\overline{\Omega}(C^\infty) + \overline{\Omega}(\Psi^{-\infty}_{\mathcal{A}\Gamma}) \to \overline{\Omega}(\Psi^0_{\mathcal{A}\Gamma})) \end{array} \tag{6.31}$$

where the vertical maps are induced by the inclusions. Moreover, by naturality also the following diagram commutes:
(6.32)

$$\begin{array}{ccc} H_{\mathrm{odd}}(\overline{\Omega}(C^\infty) + \overline{\Omega}(\Psi^{-\infty}_{\mathbb{C}\Gamma}) \to \overline{\Omega}(\Psi^0_{\mathbb{C}\Gamma})) & \xrightarrow{(\overline{\mathrm{TR}}, \overline{\mathrm{TR}}^{del})_*} & H_{\mathrm{odd}}(\Omega(\mathbb{C}\Gamma) \to \Omega^{del}(\mathbb{C}\Gamma)) \\ \downarrow & & \downarrow \\ H_{\mathrm{odd}}(\overline{\Omega}(C^\infty) + \overline{\Omega}(\Psi^{-\infty}_{\mathcal{A}\Gamma}) \to \overline{\Omega}(\Psi^0_{\mathcal{A}\Gamma})) & \xrightarrow{(\overline{\mathrm{TR}}, \overline{\mathrm{TR}}^{del})_*} & H_{\mathrm{odd}}(\widehat{\Omega}(\mathcal{A}\Gamma) \to \widehat{\Omega}^{del}(\mathcal{A}\Gamma)) \end{array}$$

$$\begin{array}{ccc} H_{\mathrm{odd}}(\Omega(\mathbb{C}\Gamma) \to \Omega^{del}(\mathbb{C}\Gamma)) & \xleftarrow[\cong]{J^{alg}_*} & H_{\mathrm{even}}(\Omega^e(\mathbb{C}\Gamma)) \\ \downarrow & & \downarrow{\scriptstyle j_*} \\ H_{\mathrm{odd}}(\widehat{\Omega}(\mathcal{A}\Gamma) \to \widehat{\Omega}^{del}(\mathcal{A}\Gamma)) & \xleftarrow[\cong]{J_*} & H_{\mathrm{even}}(\widehat{\Omega}^e(\mathcal{A}\Gamma)) \end{array} \tag{6.33}$$

where we recall that J is the inclusion of $\Omega^e(\mathcal{A}\Gamma)$ into $(\Omega(\mathcal{A}\Gamma) \to \Omega^{del}(\mathcal{A}\Gamma))$ and J^{alg} is defined similarly. Now, recall that Ch^e_Γ is defined as the composition of the bottom rows of (6.31), (6.32), and (6.33). Then, it follows that we have the following algebraic characterization of it:

$$\mathrm{Ch}^e_\Gamma = j_* \circ (J^{alg}_*)^{-1} \circ (\overline{\mathrm{TR}}, \overline{\mathrm{TR}}^{del})_* \circ \mathrm{Ch}^{rel}_{alg} \circ (i_*)^{-1}.$$

We are finally in the position of proving the following result.

PROPOSITION 6.26. *Let* $j \colon \Omega^e_*(\mathbb{C}\Gamma) \hookrightarrow \widehat{\Omega}^e_*(\mathcal{A}\Gamma)$ *be the natural inclusion. The localized Chern character factors through the algebraic one* (6.29), *specifically*

$$\mathrm{Ch}^e_\Gamma = j_* \circ \mathrm{Ch}^e_{\Gamma,alg} \colon K_0(C^\infty(M) \to \Sigma) \to H^e_{\mathrm{even}}(\mathcal{A}\Gamma). \tag{6.34}$$

Moreover, if $[e_0, e_1; \sigma] \in K_0(C^\infty(M) \to \Sigma)$ *is defined by a Dirac-type operator, then*

$$\mathrm{Ch}^e_\Gamma([e_0, e_1; \sigma]) = j_* \left[\int_M \mathrm{AS}(M, \sigma) \wedge \omega_{\mathrm{Lott}} \right] \tag{6.35}$$

with $\mathrm{AS}(M,\sigma)$ *equal to the Atiyah-Singer integrand corresponding to* σ *and with* $\omega_{\mathrm{Lott}}(M) \in \Omega^*(M) \otimes \Omega_*(\mathbb{C}\Gamma)$ *equal to Lott's bi-form.*

PROOF. We only have to prove (6.34), given that (6.35) follows directly from Proposition 6.24. The main difficulty is that $\mathrm{Ch}^e_{\Gamma,alg}$ involves taking the K-theoretic index map, whereas Ch^e_Γ doesn't. However, the commutativity of (6.30) is the commutativity of the top rectangle in the following diagram. It follows then immediately that the whole diagram below commutes:

$$\begin{array}{ccc}
K_0(C^\infty + \Psi^{-\infty}_{\mathbb{C}\Gamma} \to \Psi^0_{\mathbb{C}\Gamma}) & \xrightarrow{\cong} \; K_0(C^\infty \to \Sigma) \; \xrightarrow{\mathrm{Ind}_{\mathbb{C}\Gamma}} & K_0(\Psi^{-\infty}_{\mathbb{C}\Gamma}) \\
\Big\downarrow \mathrm{Ch}^{rel}_{alg} & & \Big\downarrow \mathrm{Ch} \\
H_{\mathrm{odd}}(\overline{\Omega}(C^\infty) + \overline{\Omega}(\Psi^{-\infty}_{\mathbb{C}\Gamma}) \to \overline{\Omega}(\Psi^0_{\mathbb{C}\Gamma})) & \xrightarrow{\delta'_c = S \circ pr} & H_{\mathrm{even}}(\overline{\Omega}(\Psi^{-\infty}_{\mathbb{C}\Gamma})) \\
\Big\downarrow (\overline{\mathrm{TR}}, \overline{\mathrm{TR}}^{del}) & & \Big\downarrow \overline{\mathrm{TR}} \\
H_{\mathrm{odd}}(\Omega(\mathbb{C}\Gamma) \to \Omega^{del}(\mathbb{C}\Gamma)) & \xrightarrow{S \circ pr} & H_{\mathrm{even}}(\Omega(\mathbb{C}\Gamma)) \\
\Big\uparrow J^{alg}_* & & \\
H_{\mathrm{even}}(\Omega^e(\mathbb{C}\Gamma)) & &
\end{array}$$

Here δ'_c is defined in (4.27), i.e. indeed is induced by the projection onto $\overline{\Omega}(\Psi^{-\infty}_{\mathbb{C}\Gamma} \to 0)$ composed with suspension to $H_{\mathrm{even}}(\overline{\Omega}(\Psi^{-\infty}_{\mathbb{C}\Gamma}))$. Consequently, the composition of the lower row is the inclusion $H_{\mathrm{even}}(\Omega^e(\mathbb{C}\Gamma)) \hookrightarrow H_{\mathrm{even}}(\Omega(\mathbb{C}\Gamma))$. Equality (6.34) then follows from Remark 6.25. □

REMARK 6.27. Since $K_0(C^\infty(M) \to \Sigma)$ is generated modulo torsion by Dirac classes and since the range of $\overline{\mathrm{TR}}_* \circ \mathrm{Ch} \circ \mathrm{Ind}_{\mathbb{C}\Gamma}$ is a $\mathbb{C}$-vector space, Proposition 6.26 gives a complete description of the homomorphism Ch^e_Γ. Moreover, we have dealt here with the even dimensional case. We expect that a formula similar to (6.35) could be obtained by suspension in the odd dimensional case.

6.5. Comparison with Lott's ϱ form

Let $\widetilde{D}_M$ be an L^2-invertible Γ-equivariant Dirac operator on a Galois Γ-covering $\widetilde{M} \to M$ acting on the sections of a Γ-equivariant complex vector bundle $\widetilde{E}$. This subsection is devoted to a comparison of our class $\mathrm{Ch}^{del}_\Gamma([\varrho(\widetilde{D}_M)])$ with the higher rho invariant of Lott.

Let $\widetilde{D}_M$ be an L^2-invertible Γ-equivariant Dirac operator on a Galois Γ-covering $\widetilde{M} \to M$ acting on the sections of a Γ-equivariant complex vector bundle $\widetilde{E}$. This subsection is devoted to a comparison of our class $\mathrm{Ch}^{del}_\Gamma([\varrho(\widetilde{D}_M)])$ with the higher rho invariant of Lott.

The higher rho invariant of Lott appeared for the first time in [**52**], as the delocalized part of its higher eta invariant. Lott's higher eta invariant and Lott's higher rho invariant were originally defined under the additional assumption that Γ is of polynomial growth (see below for the general case). Lott's higher eta invariant for Γ of polynomial growth is an element

$$\eta_{\mathrm{Lott}}(\widetilde{D}_M) \in \widehat{\Omega}_*(\mathcal{A}\Gamma)_{ab} \,,$$

with $\mathcal{A}\Gamma \subset C^*_{red}\Gamma$ denoting the algebra of rapidly decreasing functions on Γ. We shall review the definition of Lott's higher eta invariant after this brief introduction.

It was conjectured in [**52**] that the higher eta invariant was the boundary correction term in a higher Atiyah-Patodi-Singer index theorem. The conjecture was settled in [**44**]. Later, following an idea of Lott [**54**], the higher eta invariant and the higher Atiyah-Patodi-Singer index theorem were established for any finitely generated discrete group, see [**45**]. In this generality the algebra $\mathcal{A}\Gamma \subset C^*_{red}\Gamma$ can be taken to be the Connes-Moscovici algebra, using crucially that the latter is the projective limit of involutive Banach algebras with unit. Wahl [**77**] extended these results even further, allowing $\mathcal{A}\Gamma \subset C^*_{red}\Gamma$ to be *any* projective limit of involutive Banach algebras with unit. Thus, following Wahl, in the sequel we shall assume that we have a projective system of involutive Banach algebras with units $(\mathcal{A}_j, \iota_{j+1,j} : \mathcal{A}_{j+1} \to \mathcal{A}_j)_{j\in\mathbb{N}}$ satisfying the following conditions:

(i) $\mathcal{A}_0 = C^*_{red}\Gamma$;
(ii) the map $\iota_{j+1,j} : \mathcal{A}_{j+1} \to \mathcal{A}_j$ is injective for any j;
(iii) the induced map ι_j from the projective limit $\mathcal{A}_\infty$ to $\mathcal{A}_j$ has dense image;
(iv) for any $j > 0$ the algebra $\mathcal{A}_j$ is holomorphically closed in $\mathcal{A}_0 \equiv C^*_{red}\Gamma$
(v) $\mathbb{C}\Gamma$ is a subalgebra of each $\mathcal{A}_j$.

Under these assumptions we can consider $\mathcal{A}\Gamma := \mathcal{A}_\infty$, an involutive m-convex Fréchet algebra with unit. As already remarked, the Connes-Moscovici algebra $\mathcal{B}_\infty$ satisfies these hypothesis. If Γ is Gromov hyperbolic then a particular algebra $\mathcal{A}\Gamma$ defined by Puschnigg in [**68**] and discussed in detail in Section 7.3 below also satisfies these hypothesis. This ends our brief introduction to Lott's higher eta invariant. Let us move now to the actual definition.

Lott's higher eta invariant. We shall now recall the definition of Lott's higher eta invariant, a non-commutative analogue of the Bismut-Cheeger eta form. We shall follow the sign conventions given in [**51**], [**52**], based in turn on those of [**9**]. The definition of higher eta invariants is slightly different depending on the parity of the dimension. Assume that we have an even dimensional compact Riemannian manifold (M, g) and a $\mathbb{Z}_2$-graded bundle of hermitian Clifford modules, E, endowed with a Clifford unitary connection ∇^E. We denote by D the associated Dirac operator. We also have a Galois Γ-covering $\widetilde{M}$ and denote by $\tilde{g}$, $\widetilde{E}$ and $\widetilde{D}$ the corresponding equivariant lifts. We assume that $\widetilde{D}$ is L^2-invertible. For endomorphisms of $E = E_+ \oplus E_-$ we have a natural fiberwise supertrace Str. Consider now the rescaled superconnection

$$t\widetilde{D} + \nabla^{\text{Lott}} \colon \mathcal{E}^{\text{MF}}_{\mathcal{A}\Gamma}(\widetilde{M}, \widetilde{E}) \to \mathcal{E}^{\text{MF}}_{\mathcal{A}\Gamma}(\widetilde{M}, \widetilde{E}) \oplus \mathcal{E}^{\text{MF}}_{\Omega_1(\mathcal{A}\Gamma)}(\widetilde{M}, \widetilde{E})$$

Following [**52**], [**45**] and, more generally, [**77**] we can consider the superconnection heat-kernel

$$\exp(-(t\widetilde{D} + \nabla^{\text{Lott}})^2)$$

an element in $\Psi^{-\infty}_{\widehat{\Omega}_\bullet(\mathcal{A}\Gamma)}(\widetilde{M}, \widetilde{E})$ and

$$\widetilde{D}\exp(-(t\widetilde{D} + \nabla^{\text{Lott}})^2)$$

also an element in $\Psi^{-\infty}_{\widehat{\Omega}_\bullet(\mathcal{A}\Gamma)}(\widetilde{M}, \widetilde{E})$. Proceeding as in Subsection 6.2 we have a well-defined trace functional

$$\text{STR} : \Psi^{-\infty}_{\widehat{\Omega}_\bullet(\mathcal{A}\Gamma)}(\widetilde{M}, \widetilde{E}) \to \widehat{\Omega}_*(\mathcal{A}\Gamma)_{ab}$$

and the higher eta invariant of Lott is defined as

$$\eta_{\mathrm{Lott}}(\widetilde{D}_M) := \frac{2}{\sqrt{\pi}} \int_0^\infty \mathrm{STR}\left(\widetilde{D} \exp(-(t\widetilde{D} + \nabla^{\mathrm{Lott}})^2)\right) dt \quad \in \widehat{\Omega}_*(\mathcal{A}\Gamma)_{ab}. \tag{6.36}$$

where the existence of the integral is non-trivial both at 0 and at $+\infty$. (The coefficient in front comes from our choice of rescaling Lott's superconnection with t instead of $t^{\frac{1}{2}}$). We set

$$\eta_{\mathrm{Lott}}(\widetilde{D}_M)(t) := \frac{2}{\sqrt{\pi}} \mathrm{STR}\left(\widetilde{D} \exp(-(t\widetilde{D} + \nabla^{\mathrm{Lott}})^2)\right)$$

so that

$$\eta_{\mathrm{Lott}}(\widetilde{D}_M) = \int_0^\infty \eta_{\mathrm{Lott}}(\widetilde{D}_M)(t) dt \quad \in \quad \widehat{\Omega}_*(\mathcal{A}\Gamma)_{ab}$$

One can show easily, that, in fact, $\eta_{\mathrm{Lott}}(\widetilde{D}_M) \in \widehat{\Omega}_{\mathrm{odd}}(\mathcal{A}\Gamma)_{ab}$ if $\widetilde{M}$ is even-dimensional. See [**9**, Ch. 9] for the case of families, the argument is exactly the same here.

If (M, g) is odd dimensional and E is a bundle of hermitian Clifford module endowed with a Clifford unitary connection ∇^E then we proceed as follows. Denote by $\mathbb{C}\mathrm{l}(1)$ the complexified Clifford algebra associated to $\mathbb{R}$ with the standard metric; it is a unital complex algebra with generators 1 and σ and the relation $\sigma^2 = 1$. We consider $E_\sigma := E \otimes \mathbb{C}\mathrm{l}(1)$. We can identify E_σ with $E \oplus E$ and under this identification σ defines a bundle endomorphism that acts as

$$\begin{pmatrix} 0 & 1 \\ 1 & 0 \end{pmatrix}.$$

For endomorphisms of E_σ as a Clifford module we have a natural fiberwise supertrace:

$$\mathrm{Str}_\sigma(A + \sigma A') := \mathrm{Tr}(A').$$

Consider now the rescaled superconnection

$$t\sigma\widetilde{D} + \nabla^{\mathrm{Lott}} \colon \mathcal{E}^{\mathrm{MF}}_{\mathbb{C}\Gamma}(\widetilde{M}, \widetilde{E}_\sigma) \to \mathcal{E}^{\mathrm{MF}}_{\mathbb{C}\Gamma}(\widetilde{M}, \widetilde{E}_\sigma) \oplus \mathcal{E}^{\mathrm{MF}}_{\Omega_1(\mathbb{C}\Gamma)}(\widetilde{M}, \widetilde{E}_\sigma)$$

where ∇^{Lott} is a shorthand for $\nabla^{\mathrm{Lott}} \otimes \mathrm{Id}_{\mathbb{C}^2}$. We then have

$$\exp(-(t\sigma\widetilde{D} + \nabla^{\mathrm{Lott}})^2) \quad \text{and} \quad \sigma\widetilde{D}\exp(-(t\sigma\widetilde{D} + \nabla^{\mathrm{Lott}})^2),$$

both elements in $\Psi^{-\infty}_{\widehat{\Omega}_\bullet(\mathcal{A}\Gamma)}(\widetilde{M}, \widetilde{E}_\sigma)$. We have a well-defined trace functional

$$\mathrm{STR}_\sigma : \Psi^{-\infty}_{\widehat{\Omega}^+_\bullet(\mathcal{A}\Gamma)}(\widetilde{M}, \widetilde{E}_\sigma) \to \widehat{\Omega}_\bullet(\mathcal{A}\Gamma)_{ab}$$

and the higher eta invariant of Lott is defined in this odd dimensional case as

$$\begin{aligned} \eta_{\mathrm{Lott}}(\widetilde{D}_M) &:= \frac{2}{\sqrt{\pi}} \int_0^\infty \mathrm{STR}_\sigma\left(\sigma\widetilde{D}\exp(-(t\sigma\widetilde{D} + \nabla^{\mathrm{Lott}})^2)\right) dt \\ &\equiv \int_0^\infty \eta_{\mathrm{Lott}}(\widetilde{D})(t) dt \in \widehat{\Omega}_*(\mathcal{A}\Gamma)_{ab}. \end{aligned} \tag{6.37}$$

One can show that $\eta_{\mathrm{Lott}}(\widetilde{D}_M) \in \widehat{\Omega}_{\mathrm{even}}(\mathcal{A}\Gamma)_{ab}$ in the odd-dimensional case. See [**11**] for the case of families; once again, the argument is the same here.

PROPOSITION 6.28. *For the invertible operator $\widetilde{D}$ the following equality holds:*

$$d\eta_{\mathrm{Lott}}(\widetilde{D}_M) = -\int_M \mathrm{AS}(M) \wedge \omega_{\mathrm{Lott}}(M) \quad \text{in} \quad \widehat{\Omega}_*(\mathcal{A}\Gamma)_{ab}.$$

Here $\mathrm{AS}(M)$ *is the local Atiyah-Singer integrand and* $\omega_{\mathrm{Lott}}(M) \in \Omega^*(M) \otimes \Omega_*(\mathbb{C}\Gamma)$ *denotes again Lott's bi-form.*

PROOF. Assume, for example, that M is even dimensional. Then, we have the following variational formula, see [**9**], [**52**]:

$$\begin{aligned}&\mathrm{STR}\left(\exp(-(T\widetilde{D} + \nabla^{\mathrm{Lott}})^2)\right) - \mathrm{STR}\left(\exp(-(t\widetilde{D} + \nabla^{\mathrm{Lott}})^2)\right)\\ &= d\int_t^T \eta_{\mathrm{Lott}}(\widetilde{D}_M)(t)dt\,.\end{aligned}$$

Using the invertibility of the operator one can prove, see [**77**] for the most general assumptions on $\mathcal{A}\Gamma$, that

$$\lim_{T\to+\infty} \mathrm{STR}\left(\exp(-(T\widetilde{D} + \nabla^{\mathrm{Lott}})^2)\right) = 0$$

whereas Lott's theorem on the short time behaviour of the supertrace of Lott's superconnection heat-kernel, see [**51**, Proposition 12] (and [**44**, Theorem 13.6] for the version employed here), gives us

$$\lim_{t\to 0^+} \mathrm{STR}\left(\exp(-(t\widetilde{D} + \nabla^{\mathrm{Lott}})^2)\right) = \int_M \mathrm{AS}(M) \wedge \omega_{\mathrm{Lott}}(M)\,.$$

This implies the result in the even dimensional case. The odd dimensional case is similar. □

Following [**76**], we define Lott's higher rho invariant $\{\eta_{\mathrm{Lott}}(\widetilde{D})\}$ as the image of $\eta_{\mathrm{Lott}}(\widetilde{D})$ in the quotient $\widehat{\Omega}^{del}_*(\mathcal{A}\Gamma)_{ab} = \widehat{\Omega}_*(\mathcal{A}\Gamma)_{ab}/\widehat{\Omega}^e_*(\mathcal{A}\Gamma)_{ab}$. Since $\omega_{\mathrm{Lott}}(M)$ is localized at the identity element by [**51**],

$$d\{\eta_{\mathrm{Lott}}(\widetilde{D})\} = 0 \quad \text{in} \quad \widehat{\Omega}^{del}_*(\mathcal{A}\Gamma)_{ab}$$

and we obtain

$$\varrho_{\mathrm{Lott}}(\widetilde{D}) := [\{\eta_{\mathrm{Lott}}(\widetilde{D})\}] \in H^{del}_*(\mathcal{A}\Gamma)$$

with $*$ equal to even/odd if the dimension of M is odd/even. From now on we shall omit the curly brackets from the notation.

Comparing $\varrho_{\mathrm{Lott}}(\widetilde{D})$ and $\mathrm{Ch}^{del}_\Gamma([\varrho(\widetilde{D})])$. We want to compare $\varrho_{\mathrm{Lott}}(\widetilde{D})$ with our class $\mathrm{Ch}^{del}_\Gamma([\varrho(\widetilde{D})])$, which are both elements of the group $H^{del}_*(\mathcal{A}\Gamma)$. We carry out this program only in the bounding case, where we can use results from higher APS index theory, and leave the general case to future investigations.

We start with our cocompact Γ-covering $\Gamma \to \widetilde{M} \to M$ without boundary, which is endowed with a Γ-equivariant metric and with a Γ-equivariant Clifford module E. We denote as usual by $\widetilde{D}_M$ the associated Dirac-type operator. We assume that there is a cocompact Γ-covering with boundary $\Gamma \to \widetilde{W} \to W$ such that $\partial\widetilde{W} = \widetilde{M}$. We assume the existence of a Γ-equivariant metric and of a Γ-equivariant Clifford module F over $\widetilde{W}$ and denote by $\widetilde{D}_W$ the associated Dirac-type operator. If W is even-dimensional then $F = F^+ \oplus F^-$ and $\widetilde{D}_W$ is $\mathbb{Z}_2$-graded and odd:

$$\begin{pmatrix} 0 & \widetilde{D}^-_W \\ \widetilde{D}^+_W & 0 \end{pmatrix}.$$

We assume that all these structures on $\widetilde{W}$ are of product-type near the boundary and restrict to the ones of $\widetilde{M}$ on the boundary. Here we follow the conventions explained in detail in [**47**, Section 3]. Consequently, the boundary operator $\widetilde{D}^{\partial}_W$ of $\widetilde{D}_W$ is equal to $\widetilde{D}_M$; more precisely $\widetilde{D}^{+,\partial}_W = \widetilde{D}_M$ if W is even dimensional and $\widetilde{D}^{\partial}_W = \widetilde{D}_M$ if W is odd-dimensional case and with the $\mathbb{Z}_2$-grading of $\widetilde{D}_M$ given in terms of Clifford multiplication by the normal vector to the boundary, see [**47**, Section 3.2].

Our strategy is now the following: using the delocalized Chern character, we wish to connect the delocalized APS index theorem in K-theory, proved in [**65**], with the delocalized APS index theorem in non-commutative de Rham homology. The latter is obtained from the following higher index formula, proved in [**45**, Theorem 4.1], [**46**, (3.1)], [**77**, Theorem 9.4, Theorem 11.1] and stated below for the benefit of the reader:

THEOREM 6.29. *Under the above assumptions, there exists a well defined APS index class* $\mathrm{Ind}_{\Gamma,b}(\widetilde{D}_W) \in K_*(\mathcal{A}\Gamma)$ *such that*
(6.38)
$$\mathrm{Ch}_\Gamma(\mathrm{Ind}_{\Gamma,b}(\widetilde{D}_W)) = \left[\int_W \mathrm{AS}(W) \wedge \omega_{\mathrm{Lott}}(W) - \frac{1}{2}\eta_{\mathrm{Lott}}(\widetilde{D}_M)\right] \quad \in H_*(\mathcal{A}\Gamma).$$
In this formula, $\omega_{\mathrm{Lott}}(W)$ *is again the explicit bi-form due to Lott,* [**51**] [**44**, Theorem 13.6]; *we recall that* $\omega_{\mathrm{Lott}}(W)$ *is concentrated at the identity element of the group.*

REMARK 6.30. The index class $\mathrm{Ind}_{\Gamma,b}(\widetilde{D}_W) \in K_*(\mathcal{A}\Gamma)$ is associated to the Dirac operator defined by $\widetilde{D}_W$ on the Γ-covering with cylindrical ends associated to $\widetilde{W}$. In the framework on Melrose, this is a b-manifold, hence the subscript. There is also an APS-index class $\mathrm{Ind}_{\Gamma,\mathrm{APS}}(\widetilde{D}_W) \in K_*(\mathcal{A}\Gamma)$ defined via an APS-boundary value problem and one can prove [**47**], as in the numeric case treated by Atiyah, Patodi and Singer, that $\mathrm{Ind}_{\Gamma,b}(\widetilde{D}_W) = \mathrm{Ind}_{\Gamma,\mathrm{APS}}(\widetilde{D}_W)$ in $K_*(\mathcal{A}\Gamma)$. With a small and well-established abuse of terminology we refer to this index class as a higher APS-index class.

By applying $q_*\colon H_*(\mathcal{A}\Gamma) \to H^{del}_*(\mathcal{A}\Gamma)$ to both sides and using that $\omega_{\mathrm{Lott}}(W)$ is concentrated at the identity element, we obtain the

PROPOSITION 6.31 (Homological delocalized APS index theorem using superconnections). *The following equality holds:*

$$q_*(\mathrm{Ch}_\Gamma(\mathrm{Ind}_{\Gamma,b}(\widetilde{D}_W))) = -\frac{1}{2}\varrho_{\mathrm{Lott}}(\widetilde{D}_M) \quad \in H^{del}_*(\mathcal{A}\Gamma)\,. \tag{6.39}$$

Let us go back to our task of comparing $\varrho_{\mathrm{Lott}}(\widetilde{D}_M)$ with $\mathrm{Ch}^{del}_\Gamma(\varrho(\widetilde{D}_M))$. If we could prove that

$$q_*(\mathrm{Ch}_\Gamma(\mathrm{Ind}_{\Gamma,b}(\widetilde{D}_W))) = \mathrm{Ch}^{del}_\Gamma(\varrho(\widetilde{D}_M)) \quad \in H^{del}_*(\mathcal{A}\Gamma) \tag{6.40}$$

then, clearly, we would obtain the equality

$$\mathrm{Ch}^{del}_\Gamma([\varrho(\widetilde{D}_M)]) = -\frac{1}{2}\varrho_{\mathrm{Lott}}(\widetilde{D}_M) \quad \in H^{del}_*(\mathcal{A}\Gamma).$$

We shall explain this result in the even dimensional:

PROPOSITION 6.32 (Homological delocalized APS index theorem using the delocalized Chern character). *Assume that* $\widetilde{W}$ *is even dimensional. Then the following formula holds*

$$q_*(\mathrm{Ch}_\Gamma(\mathrm{Ind}_{\Gamma,b}(\widetilde{D}_W))) = \mathrm{Ch}^{del}_\Gamma(\varrho(\widetilde{D}_M)) \quad \in H^{del}_{\mathrm{even}}(\mathcal{A}\Gamma). \tag{6.41}$$

We postpone the proof of this proposition to the next subsection.

Summarizing, from (6.39) and (6.41) we have obtained the following result:

PROPOSITION 6.33. *If $\widetilde{M}$ is odd dimensional, $\widetilde{M} = \partial \widetilde{W}$ and $\widetilde{D}_M = \widetilde{D}_W^{+,\partial}$ as above, then*

$$\mathrm{Ch}_\Gamma^{del}([\varrho(\widetilde{D}_M)]) = -\frac{1}{2}\varrho_{\mathrm{Lott}}(\widetilde{D}_M) \quad \in H^{del}_{\mathrm{even}}(\mathcal{A}\Gamma).$$

At least in the bounding case this answers a question raised by Lott in [**52**, Remark 4.11.3]. We leave the general non-bounding case to future investigations; we expect the techniques in [**25**, Section 5] to play a crucial role.

6.6. Proof of Proposition 6.32

We would like to obtain (6.41) by employing the delocalized APS index theorem in K-Theory proved in [**65**]. However, since we are using a different description of the analytic surgery sequence, we shall first need to reformulate the delocalized APS index theorem in K-theory in our new setting. To this end, we recall and slightly modify results due to the third author, see [**90**]. We shall freely use Lie groupoids techniques; we refer the reader, for example, to [**67**, Section 4, Section 8.1, 8.2] for a quick introduction to the relevant definitions and basics results.

We start by observing that in the above geometric setting we have the following b-groupoid,

$$\widetilde{\mathring{W}} \times_\Gamma \widetilde{\mathring{W}} \cup \widetilde{M} \times_\Gamma \widetilde{M} \times \mathbb{R} \rightrightarrows W.$$

The smooth structure of this Lie groupoid is rigorously given by the fact that it is the blow-up of $\widetilde{M} \times_\Gamma \widetilde{M}$ into $\widetilde{W} \times_\Gamma \widetilde{W}$, see for instance [**22**]. In order to lighten the notation we will denote by $G(M)$ the groupoid $\widetilde{M} \times_\Gamma \widetilde{M}$, by $G(W, M)$ the b-groupoid, and by $G(\mathring{W})$ the groupoid $\widetilde{\mathring{W}} \times_\Gamma \widetilde{\mathring{W}}$.

We have the following commutative diagram of K-theory groups

(6.42)

$$\begin{array}{ccccc}
K_0(C^*_{red}(G(\mathring{W}))) & \xrightarrow[\cong]{S} & K_1(C^*_{red}(G(\mathring{W})) \otimes C_0(0,1)) & \overset{i_*}{\searrow} & \\
\Big\downarrow j_* & & \Big\downarrow j^c_* & & \\
K_0(\Psi^0_\Gamma(\widetilde{\mathring{W}})) & \longrightarrow & K_1(C_0(\mathring{W}) \to \Psi^0_\Gamma(\widetilde{\mathring{W}})) & \xrightarrow{\cong} & K_1(C^*_{red}(G(\mathring{W})^{[0,1)}_{ad})) \\
\Big\uparrow \partial^\Psi & & \Big\uparrow \partial^c & & \Big\uparrow \partial^{ad} \\
K_1(\Psi^0_{\Gamma,\mathbb{R}}(\widetilde{M} \times \mathbb{R})) & \longrightarrow & K_0(C(M) \to \Psi^0_{\Gamma,\mathbb{R}}(\widetilde{M} \times \mathbb{R})) & \xrightarrow{\cong} & K_0(C^*_{red}((G(M) \times \mathbb{R})^{[0,1)}_{ad})) \\
\Big\uparrow s^\Psi & & & & \\
K_0(\Psi^0_\Gamma(\widetilde{M})) & & & &
\end{array}$$

In this diagram:

- S denotes the suspension isomorphism;
- i is the natural inclusion of $C^*_{red}(G(\mathring{W})) \otimes C_0(0,1)$ into $C^*_{red}(G(\mathring{W})^{[0,1)}_{ad})$;

- j and j^c are the natural inclusions of $C^*_{red}(G(\mathring{W}))$ into $\Psi^0_\Gamma(\widetilde{\mathring{W}})$ and of $C^*_{red}(G(\mathring{W})\otimes C_0(0,1))$ into the mapping cone of $C_0(\mathring{W}) \to \Psi^0_\Gamma(\widetilde{\mathring{W}})$ respectively, the horizontal isomorphisms on the right hand side are given by equation (2.11) in [**89**].

Next we define the homomorphisms ∂^Ψ, ∂^c, ∂^{ad} and S^Ψ, thus completing the description of the above diagram.

For the homomorphisms ∂^Ψ, we consider $\Psi^0_{\Gamma,b}(\widetilde{W})$, the C*-closure of the 0-order Γ-equivariant b-pseudodifferential operators on $\widetilde{W}$ of Γ-compact support, which is the same as the C*-closure of the compactly supported 0-order pseudodifferential operators on $G(W,M)$. Then the restriction to the boundary gives rise to the following short exact sequences

$$0 \longrightarrow \Psi^0_\Gamma(\widetilde{\mathring{W}}) \longrightarrow \Psi^0_{\Gamma,b}(\widetilde{W}) \longrightarrow \Psi^0_{\Gamma,\mathbb{R}}(\widetilde{M}\times\mathbb{R}) \longrightarrow 0, \tag{6.43}$$

where $\Psi^0_{\Gamma,\mathbb{R}}(\widetilde{M}\times\mathbb{R})$ are the suspended operators on $\widetilde{M}$, namely operators on $\widetilde{M}\times\mathbb{R}$ which are translation invariant on $\mathbb{R}$. Then ∂^Ψ is the boundary morphism associated to (6.43). Analogously, one defines ∂^c by means of the mapping cone construction.

For the homomorphism ∂^{ad}, we consider the boundary map associated to the following short exact sequence

$$0 \longrightarrow C^*_{red}(G(\mathring{W})^{[0,1)}_{ad}) \longrightarrow C^*_{red}(G(W,M)^{[0,1)}_{ad}) \longrightarrow C^*_{red}((G(M)\times\mathbb{R})^{[0,1)}_{ad}) \longrightarrow 0, \tag{6.44}$$

where we recall that the adiabatic deformation of a Lie groupoid $G \rightrightarrows G^{(0)}$ is given by $\mathfrak{A}G\times\{0\}\sqcup G\times(0,1) \rightrightarrows G^{(0)}\times[0,1)$, with $\mathfrak{A}G$ being the Lie algebroid of G, see [**22,67**] where the details of these constructions are given. Recall that the K-theory of $C^*_{red}(G(M)^{[0,1)}_{ad})$ gives another realization of the structure group $\mathrm{S}^\Gamma_*(\widetilde{M})$, see [**89**].

Finally, the map S^Ψ in (6.42) is given by the composition of the suspension isomorphism S followed by the homomorphism ι_* induced by the inclusion $\iota\colon \Psi^0_\Gamma(\widetilde{M})\otimes C_0(\mathbb{R}) \to \Psi^0_{\Gamma,\mathbb{R}}(\widetilde{M}\times\mathbb{R})$ given by Fourier transform in the $\mathbb{R}$ direction.

Let us now consider the Γ-equivariant Dirac b-operator $\widetilde{D}_W$ on $\widetilde{W}$. Since $\widetilde{D}_M$, the operator on the boundary, is L^2-invertible we know that $\widetilde{D}_W$ is fully elliptic and there is a well-defined b-index class $\mathrm{Ind}_{\Gamma,b}(\widetilde{D}_W)\in K_0(C^*_{red}(G(\mathring{W})))$. This index class precisely corresponds to the one appearing in (6.38) through the isomorphisms $K_0(C^*_{red}(G(\mathring{W})))\cong K_0(C^*_{red}\Gamma)\cong K_0(\mathcal{A}\Gamma)$.

In this context, the third author of this paper [**90**, Equation (3.3)] has defined the adiabatic rho class $\varrho^{ad}(\widetilde{D}_M)$ as an element in $K_0(C^*_{red}(G(M)^{[0,1)}_{ad}))$. The delocalized APS index theorem (in K-theory) in the groupoid framework, [**90**, Theorem 3.7], states that

$$i_*(S(\mathrm{Ind}_{\Gamma,b}(\widetilde{D}_W))) = \partial^{ad}(\varrho^{ad}(\widetilde{D}_M)). \tag{6.45}$$

We want to use this fundamental equality in order to prove the identity (6.41). To this end we consider the b-groupoid $G([0,1),\{0\})$. We have the following commutative diagram:

(6.46)

$$\begin{array}{ccccc} \Psi^0_\Gamma(\widetilde{M})\otimes C^*_{red}((0,1)\times(0,1)) & \hookrightarrow & \Psi^0_\Gamma(\widetilde{M})\otimes C^*_{red}G([0,1),\{0\}) & \twoheadrightarrow & \Psi^0_\Gamma(\widetilde{M})\otimes C_0(\mathbb{R}) \\ \downarrow{\scriptstyle \iota'} & & \downarrow{\scriptstyle \iota''} & & \downarrow{\scriptstyle \iota} \\ \Psi^0_\Gamma(\overset{\circ}{\widetilde{W}}) & \hookrightarrow & \Psi^0_{\Gamma,b}(\widetilde{W}) & \twoheadrightarrow & \Psi^0_{\Gamma,\mathbb{R}}(\widetilde{M}\times\mathbb{R}) \end{array}$$

with ι from above and where ι'' is defined by the fact that over a collar neighbourhood of the boundary $G(W,M)$ is isomorphic to $G(M)\times G([0,1),\{0\})$. Finally, ι' is the restriction of ι''. We denote by $\mathrm{id}\otimes\partial$ the K-theory boundary map of the first row. Notice that ∂ is the isomorphism which sends the generator of $K_1(C_0(\mathbb{R}))$ to the generator of $K_0(C^*_{red}((0,1)\times(0,1)))$, namely the class of a rank one projector $e\in\mathbb{K}(L^2(0,1))$. Now, consider $x\in K_0(\Psi^0_\Gamma(\widetilde{M}))$, then by naturality and by the above remarks

$$\partial^\Psi(S^\Psi(x))=\partial^\Psi(\iota_*(Sx))=\iota'_*(\mathrm{id}\otimes\partial(Sx))=\iota'_*(x\otimes e). \tag{6.47}$$

Consider now the class $[\pi_\geq(\widetilde{D}_M)]=:[\pi_\geq]\in K_0(\Psi^0_\Gamma(\widetilde{M}))$ given by the projection on the positive spectrum of the Dirac operator $\widetilde{D}_M$ on $\widetilde{M}$. We know from [**89**, Section 5.3] that $[\pi_\geq]$ is sent to the adiabatic rho class $\varrho^{ad}(\widetilde{D}_M)\in K_0(C^*_{red}(G(M)^{[0,1)}_{ad}))$ by the composition of the homomorphisms appearing in the bottom row in (6.42). By using the delocalized APS index theorem in the groupoid framework, i.e. the equality $i_*(S(\mathrm{Ind}_{\Gamma,b}(\widetilde{D}_W)))=\partial^{ad}(\varrho^{ad}(\widetilde{D}_M))$, and by a simple diagram chase in (6.42), we see that the difference

$$j_*(\mathrm{Ind}_{\Gamma,b}(\widetilde{D}_W))-\partial^\Psi(S^\Psi([\pi_\geq]))$$

is in the image of $K_0(C_0(\overset{\circ}{W}))$ and is therefore local in the sense that its Chern character lies in the image of $H^e(\mathcal{A}\Gamma)$.

We then have the following sequence of equalities

$$\begin{aligned} q_*(\mathrm{Ch}_\Gamma(\mathrm{Ind}_{\Gamma,b}(\widetilde{D}_W)))&=\mathrm{Ch}^{del}_\Gamma(j_*(\mathrm{Ind}_{\Gamma,b}(\widetilde{D}_W)))\\ &=\mathrm{Ch}^{del}_\Gamma(\partial^\Psi(S^\Psi([\pi_\geq])))\\ &=\mathrm{Ch}^{del}_\Gamma(\iota'([\pi_\geq]\otimes e))\\ &=\mathrm{Ch}^{del}_\Gamma([\pi_\geq]\otimes e)\\ &=\mathrm{Ch}^{del}_\Gamma([\pi_\geq]) \end{aligned} \tag{6.48}$$

where the first equality is given by our main Theorem 6.24; the second equality employs the fact that Ch^{del}_Γ is zero on local terms; the third equality is given by (6.47); the fourth equality is a direct consequence of the definition of ι' and finally the last equality is given by the fact that the trace of the tensor product is the product of the traces and that the trace of e is 1. Summarizing:

$$q_*(\mathrm{Ch}_\Gamma(\mathrm{Ind}_{\Gamma,b}(\widetilde{D}_W)))=\mathrm{Ch}^{del}_\Gamma([\pi_\geq]) \tag{6.49}$$

which is the equality we wanted to show.

CHAPTER 7

Higher rho numbers associated to elements in $HC^*(\mathbb{C}\Gamma, \langle x \rangle)$

7.1. The role of cyclic group cohomology

The main result of Chapter 6, and indeed the key achievement of this paper, is a compatible Chern character map from the analytic K-theory sequence to non-commutative de Rham homology sequences. These de Rham homology sequences are obtained for suitable completions $\mathcal{A}\Gamma$ of the group ring $\mathbb{C}\Gamma$ of the underlying group. To make use of this Chern character, the next step is then to obtain pairings between this homology and suitable cohomology groups. Here, one will typically rely on (periodic) cyclic cohomology.

We shall use the basic fact that the (universal) (reduced) de Rham homology of a Frèchet algebra A is indeed canonically a summand of the cyclic homology: $\overline{H}_*(A) \hookrightarrow \overline{HC}_*(A)$, [**50**, Theorem 2.6.7] or [**40**, Théorème 1.15]. We will discuss this relation and the meaning of "reduced" in more detail in Section 7.5. The cyclic homology can be defined in several canonically isomorphic ways, e.g. as the homology of the cyclic chain complex $CC_\bullet(A)$. We will give more details about the cyclic (co)chain complexes below.

By definition, cyclic cohomology pairs with cyclic homology and therefore also with non-commutative de Rham homology. In fact, the cyclic cochain complex $CC^\bullet(A)$ is the dual of the cyclic chain complex $CC_\bullet(A)$. Correspondingly, we have the pairing $CC_n(A)\otimes CC^n(A) \to \mathbb{C}$ descending to a pairing $HC_n(A)\otimes HC^n(A) \to \mathbb{C}$ on the level of cyclic (co)homology. Of course, when working with Frèchet algebras, throughout the topology has to be taken into account (topological duals, projective tensor products,...).

For a completion $\mathcal{A}\Gamma$ that one has to use for the Chern character in Chapter 6, it will in general be impossible to get direct information about $HC^*(\mathcal{A}\Gamma)$. Instead, one would like to use the pretty well understood $HC^*(\mathbb{C}\Gamma)$, i.e. one would like to use (algebraic) cyclic cocycles of the group Γ. In general, this will not be possible. The inclusion $\mathbb{C}\Gamma \hookrightarrow \mathcal{A}\Gamma$ induces a restriction map $CC^\bullet(\mathcal{A}\Gamma) \to CC^\bullet(\mathbb{C}\Gamma)$. Due to the fact that $\mathbb{C}\Gamma$ is dense in $\mathcal{A}\Gamma$ and that the cochains in $CC^\bullet(\mathcal{A}\Gamma)$ are required to be continuous, the restriction map is an injection $CC^\bullet(\mathcal{A}\Gamma) \hookrightarrow CC^\bullet(\mathbb{C}\Gamma)$. A common theme is the "extension" of cocycles from $\mathbb{C}\Gamma$ to $\mathcal{A}\Gamma$. If this is possible for all of $CC^\bullet(\mathbb{C}\Gamma)$ it simply means that we have an isomorphism $CC^\bullet(\mathcal{A}\Gamma) \xrightarrow{\cong} CC^\bullet(\mathbb{C}\Gamma)$. If this is possible for a subcomplex $E^* \subset CC^\bullet(\mathbb{C}\Gamma)$, namely we get an isomorphic subcomplex $\overline{E}^* \subset CC^\bullet(\mathcal{A}\Gamma)$ of extended cochains, then there is a well defined map $H^*(E) \xrightarrow{\cong} H^*(\overline{E}^*) \to HC^*(\mathcal{A}\Gamma)$. This fact holds for the reduced (co)homology groups, too. In our situation, E^* will be a direct summand of $CC^\bullet(\mathbb{C}\Gamma)$ and consequently $\overline{H}^*(E^*)$ will be a direct summand of $\overline{HC}^*(\mathbb{C}\Gamma)$ mapping to $\overline{HC}^*(\mathcal{A}\Gamma)$

and thus pairing with non-commutative de Rham homology. As a consequence, one can just use the cyclic cohomology of $\mathbb{C}\Gamma$ (or rather the summand $\overline{H}^*(E^*)$) to obtain pairings with $\overline{H}_*(\mathcal{A}\Gamma)$ in place of the a priori mysterious $\overline{HC}^*(\mathcal{A}\Gamma)$. This is what we will achieve for hyperbolic groups (and a convenient completion $\mathcal{A}\Gamma$ due to Puschnigg) in Chapter 7.3. More directly, this will also be achieved for groups of polynomial growth.

7.2. Cyclic cohomology of hyperbolic groups

A central role in this article will be played by the following well-known theorem.

THEOREM 7.1 (Burghelea [**13**]). *For any discrete group* Γ *we have the following isomorphisms for the Hochschild and the cyclic cohomology groups, respectively:*

(1) $HH_*(\mathbb{C}\Gamma) \cong \bigoplus_{\langle\gamma\rangle\in\langle\Gamma\rangle} H_*(\Gamma_\gamma;\mathbb{C})$,

(2) $HC_*(\mathbb{C}\Gamma) \cong \bigoplus_{\langle\gamma\rangle\in\langle\Gamma\rangle^\infty} H_*(\Gamma_\gamma/\gamma^{\mathbb{Z}};\mathbb{C}) \oplus \bigoplus_{\langle\gamma\rangle\in\langle\Gamma\rangle^{fin}} H_*(\Gamma_\gamma/\gamma^{\mathbb{Z}};\mathbb{C})\otimes\mathbb{C}[z]$.

Here, $\langle\Gamma\rangle$ *is the set of all conjugacy classes in* Γ, *and* $\langle\gamma\rangle$ *denotes the conjugacy class of* γ; $\langle\Gamma\rangle^\infty$ *(respectively* $\langle\Gamma\rangle^{fin}$*) denotes the set of conjugacy classes of elements of infinite (respectively finite) order. Moreover* $\Gamma_\gamma := \{g \in \Gamma \mid g\gamma = \gamma g\}$ *is the centralizer of* γ *in* Γ *and* $\gamma^{\mathbb{Z}}$ *denotes the cyclic group generated by* $\gamma \in \Gamma$. *Finally,* $H_*(\Gamma_\gamma;\mathbb{C})$ *is the usual group cohomology with complex coefficients, and the polynomial generator* z *is of degree* 2.

There is an analogous version of this theorem for the Hochschild and the cyclic cohomology groups, where direct sums are replaced by direct products. In what follows we shall see aspects of the proof in order to show that, under suitable assumptions on Γ, we can bring in polynomial growth conditions.

First, let us recall the definition of the cyclic set $Z\Gamma$. For more background on what follows see for instance [**78**, Section 9.7]. Here $Z_n\Gamma := \Gamma^{n+1}$ and the degeneracies, the face maps, and the cyclic structures are defined in the following way:

$$
\begin{aligned}
\partial_i(g_0,\dots,g_n) &:= \begin{cases}(g_0,\dots,g_ig_{i+1},\dots,g_n), & \text{for } i<n\\ (g_ng_0,g_1,\dots,g_{n-1}), & \text{for } i=n\end{cases}\\
\sigma_i(g_0,\dots,g_n) &:= (g_0,\dots,g_i,e,g_{i+1},\dots,g_n),\\
t(g_0,\dots,g_n) &:= (g_n,g_0,\dots,g_{n-1}).
\end{aligned} \tag{7.1}
$$

Now, for $x\in\Gamma$, let $Z_n(\Gamma,x)$ denote the subset of $Z_n\Gamma$ consisting of all $(g_0,\dots,g_n)$ such that the product $g_0\cdots g_n$ is conjugate to x. As n varies, these subsets form a cyclic subset $Z(\Gamma,x)$ of $Z\Gamma$.

The Hochschild cohomology $HH^*(\mathbb{C}\Gamma)$ is given by the cohomology of the cochain complex $C^\bullet(Z\Gamma)$ associated to the simplicial object underlying $Z\Gamma$. In particular, we have that

$$HH^*(\mathbb{C}\Gamma) = \prod_{\langle x\rangle\in\langle\Gamma\rangle} HH^*(\mathbb{C}\Gamma;\langle x\rangle) \tag{7.2}$$

which corresponds by definition to the decomposition

$$H^*(Z\Gamma) = \prod_{\langle x\rangle\in\langle\Gamma\rangle} H^*(Z(\Gamma,x)).$$

DEFINITION 7.2. Let $l\colon \Gamma \to [0,+\infty)$ be the word-length function on Γ associated to a symmetric set of generators S. Define $C^{\bullet}_{pol}(\mathbb{C}\Gamma;\langle x\rangle)$ as the subcomplex of the Hochschild complex $C^{\bullet}(\mathbb{C}\Gamma;\langle x\rangle)$ associated to $Z(\Gamma,x)$ whose elements are cochains of polynomial growth, i.e. functions $f\colon Z_n(\Gamma,x)\to\mathbb{C}$ with $|f(g_0,\dots,g_n)|\le C(1+l(g_0)\cdots l(g_n))^N$ for suitable C, N. Denote by $HH^*_{pol}(\mathbb{C}\Gamma;\langle x\rangle)$ its cohomology.

By [**78**, Proposition 9.7.4], for all $x\in\Gamma$ the inclusion $\Gamma_x\hookrightarrow\Gamma$ induces a homotopy equivalence $\iota\colon Z(\Gamma_x,x)\to Z(\Gamma,x)$ of cyclic sets with homotopy inverse $\rho\colon Z(\Gamma,x)\to Z(\Gamma_x,x)$, explicitly defined in the following way. First, choose representatives of minimal word length for the right Γ_x-coset in Γ. Given $(g_0,\dots,g_n)\in Z_n(G,x)$, let y_i be the chosen coset representative such that

$$y_i(g_{i+1}\cdots g_n g_0\cdots g_i)y_i^{-1}=x$$

and set

$$\rho(g_0,\dots,g_n):=(y_n g_0 y_0^{-1}, y_0 g_1 y_1^{-1},\dots,y_{n-1}g_n y_n^{-1}). \tag{7.3}$$

Moreover, the simplicial homotopy between the identity map of $Z(\Gamma,x)$ and $\iota\circ\rho$ is given by

(7.4)

$$h_j(g_0,\dots,g_n):=(g_0y_0^{-1}, y_0g_1y_1^{-1},\dots,y_{j-1}g_jy_j^{-1},y_j,g_j,\dots,g_n);\quad j\in\{0,\dots,n\}.$$

REMARK 7.3. Observe that the simplicial cochain complex associated to $Z(\Gamma_x,x)$ is exactly the standard bar complex which defines the group cohomology $H^*(\Gamma_x;\mathbb{C})$. Hence we have that the group cohomology $H^*(\Gamma_x;\mathbb{C})$ is the simplicial set cohomology $H^*(Z(\Gamma_x,x))$, which in turn is isomorphic to $HH^*(\mathbb{C}\Gamma;\langle x\rangle):=H^*(Z(\Gamma,x))$ by means of the pull-back map ρ^*.

Let us recall the following fundamental definition.

DEFINITION 7.4. A geodesic metric space (X,d) is δ-hyperbolic if each edge of a geodesic triangle in X is contained in the δ-neighbourhood of the union of the two other edges. Thus, if $[a,b],[b,c],[c,a]$ are geodesic segments in X, then

$$x\in[a,b]\Rightarrow d_X(x,[b,c]\cup[c,a])\le\delta.$$

It is simply called hyperbolic if it is δ-hyperbolic for some $\delta\ge 0$.

Let us assume from now on that Γ is hyperbolic, namely, by definition, that there exists a δ such that the Cayley-graph $G(\Gamma,S)$ is δ-hyperbolic. A result of Gromov implies that also Γ_x is hyperbolic for all $x\in\Gamma$, see [**27**, Section 8.5.M].

For every element $\gamma\in\Gamma$ we choose a word $w(\gamma)$ of minimal length representing it and fix an element $\sigma(\gamma)$ of minimal word length $l(\gamma)$ in the conjugacy class $\langle\gamma\rangle$. We recall the statement of [**68**, Lemma 4.1].

LEMMA 7.5. *Let Γ be hyperbolic. For given $R>0$ there exists a constant $C(R)>0$ such that the following holds. Let $\gamma\in\Gamma$ be written as a word $w(\gamma)$ in the generators S and assume that $\min_{g\in\langle\gamma\rangle} l(g)\le R$. Then some cyclic permutation of $w(\gamma)$ represents an element of length less than $C(R)$ in Γ.*

As a consequence, we have the following result.

LEMMA 7.6. *If Γ is hyperbolic and $x\in\Gamma$, the pull-back of cochains through ρ sends cochains of polynomial growth on $Z(\Gamma_x,x)$ to cochains of polynomial growth on $Z(\Gamma,x)$. The same is true for the simplicial homotopy $\{h_j\}$.*

PROOF. Start with $g_0, \dots, g_n \in \Gamma$ such that $g_0 \cdots g_n \in \langle x \rangle$. Abbreviate $G_i := g_{i+1} \cdots g_n g_0 \cdots g_i$. The construction of ρ and of the simplicial homotopy is based on the elements $y_0, \dots, y_n$ which depends on $g_0, \dots, g_n$. We show that their length is bounded linearly in the length of the g_i, which immediately implies the statement.

First, observe that each G_i is conjugated to x. Using Lemma 7.5, there is $C > 0$ (depending only on $l(x)$) such that we find a subword a_i of G_i and $l(a_i G_i a_i^{-1}) \leq C$. Note that, as a subword of a product of the g_i, $l(a_i) \leq \sum_{k=0}^n l(g_k)$. Secondly, the set of elements of $\langle x \rangle$ of length $\leq C$ is finite. Therefore, there is $D > 0$ depending only on C and x and b_i with $l(b_i) \leq D$ and such that $b_i a_i G_i a_i^{-1} b_i^{-1} = x$. Now, also $y_i G_i y_i^{-1} = x$, therefore $b_i a_i$ and y_i belong to the same Γ_x-coset. By minimality,

$$l(y_i) \leq l(b_i a_i) \leq D + \sum_{k=0}^n l(g_i).$$

Let now $f \colon \Gamma_x^{n+1} \to \mathbb{C}$ be a cochain of polynomial growth, i.e. $|f(g_0, g_1, \dots, g_n)| \leq C(1 + l(g_0) \cdots l(g_n))^N$ for some C, N. It follows now immediately from the definition of ρ that also $\rho^* f$ has polynomial growth, as

$$\begin{aligned} |\rho^* f(g_0, \dots, g_n)| &= |f(y_n g_0 y_0^{-1}, \dots, y_{n-1} g_n y_n^{-1})| \\ &\leq C(1 + l(y_n g_0 y_0^{-1}) \cdots l(y_{n-1} g_n y_n^{-1}))^N \\ &\leq C(1 + (l(y_n) + l(g_0) + l(y_0)) \cdots (l(y_{n-1}) + l(g_n) + l(y_n)))^N \end{aligned}$$

which, using our preparation, is bounded by a polynomial in $l(g_0), \dots, l(g_n)$ (even of unchanged degree). The same argument applies to the chain homotopy and the result follows. □

PROPOSITION 7.7. *Let Γ be a hyperbolic group or assume that Γ has polynomial growth, $x \in \Gamma$. Then the inclusion of complexes $C^\bullet_{pol}(\mathbb{C}\Gamma; \langle x \rangle) \hookrightarrow C^\bullet(\mathbb{C}\Gamma; \langle x \rangle)$ induces an isomorphism $HH^*_{pol}(\mathbb{C}\Gamma; \langle x \rangle) \cong HH^*(\mathbb{C}\Gamma; \langle x \rangle)$.*

PROOF. Observe preliminarily that Lemma 7.6 holds also for groups of polynomial growth, with an easier, immediate proof. If Γ is hyperbolic, the same is true for Γ_x by [**27**, Section 8.5.M]. If Γ has polynomial growth, every subgroup, in particular Γ_x also has polynomial growth. If Γ is hyperbolic it admits a combing of polynomial growth and by [**60**, Corollary 5.3] the natural inclusion of complexes $C^\bullet_{pol}(\Gamma_x; \mathbb{C}) \hookrightarrow C^\bullet(\Gamma_x; \mathbb{C})$ induces an isomorphism between $H^*_{pol}(\Gamma_x; \mathbb{C})$ and $H^*(\Gamma_x; \mathbb{C})$; here $C^\bullet_{pol}(\Gamma_x; \mathbb{C})$ is the subcomplex of the bar complex $C^\bullet(\Gamma_x; \mathbb{C})$ whose elements are cochains with polynomial growth on Γ_x. The same applies by [**59**] if Γ has polynomial growth. Now, by Lemma 7.6 ρ^* preserve the polynomial growth of cochains and we have the following commutative square

$$\begin{array}{ccc} H^*_{pol}(\Gamma_x; \mathbb{C}) & \xrightarrow{\cong} & H^*(\Gamma_x; \mathbb{C}) \\ \downarrow{\scriptstyle \rho^*} & & \downarrow{\scriptstyle \rho^*} \\ HH^*_{pol}(\mathbb{C}\Gamma; \langle x \rangle) & \longrightarrow & HH^*(\mathbb{C}\Gamma; \langle x \rangle). \end{array} \tag{7.5}$$

Since the simplicial homotopy $\{h_j\}$ preserves polynomial growth, ρ^* induces an isomorphism on both sides of the square and the result follows. □

So far we only treated Hochschild cohomology, which was constructed from the simplicial structure of $Z\Gamma$. Thanks to the cyclic structure (7.1) of $Z\Gamma$, one also obtains a bicomplex $CC^\bullet(\mathbb{C}\Gamma)$ whose total complex gives the cyclic cohomology of $\mathbb{C}\Gamma$. Moreover the k-th column of $CC^\bullet(\mathbb{C}\Gamma)$, for k even, is equal to the Hochschild complex. For more background about this see for instance [**78**, Sections 9.6, 9.7]. Let us then consider the following short exact sequence of bicomplexes

$$0 \longrightarrow CC^\bullet_{\{2\}}(\mathbb{C}\Gamma) \longrightarrow CC^\bullet(\mathbb{C}\Gamma) \longrightarrow CC^\bullet(\mathbb{C}\Gamma)[2,0] \longrightarrow 0$$

where we denote by $CC^\bullet_{\{2\}}(\mathbb{C}\Gamma)$ the bicomplex given by the first two columns of $CC^\bullet(\mathbb{C}\Gamma)$ and $[2,0]$ denotes the 2-degree horizontal shifting of the bicomplex. It turns out that $CC^\bullet_{\{2\}}(\mathbb{C}\Gamma)$ is quasi-isomorphic to the Hochschild complex. Now, taking the associated long exact sequence of the total complexes we obtain the Connes periodicity exact sequence

$$\to HH^n(\mathbb{C}\Gamma) \xrightarrow{B} HC^{n-1}(\mathbb{C}\Gamma) \xrightarrow{S} HC^{n+1}(\mathbb{C}\Gamma) \xrightarrow{I} HH^{n+1}(\mathbb{C}\Gamma) \xrightarrow{B} \tag{7.6}$$

see [**50**, Theorem 2.2.1] for the proof of the homological version of this fact.

Observe now that, since $Z\Gamma = \bigcup Z(\Gamma, x)$ is a disjoint union of cyclic subset, we have that also $CC^\bullet(\mathbb{C}\Gamma)$ decomposes as $\prod_{\langle x\rangle \in \langle \Gamma\rangle} CC^\bullet(\mathbb{C}\Gamma; \langle x\rangle)$. This decomposition is compatible with the construction of (7.6).

THEOREM 7.8. *Let the group Γ be hyperbolic or of polynomial growth, $x \in \Gamma$. Let $CC^\bullet_{pol}(\mathbb{C}\Gamma; \langle x\rangle)$ be the subcomplex of $CC^\bullet(\mathbb{C}\Gamma; \langle x\rangle)$ whose elements are cochains of polynomial growth. Then the natural inclusion induces an isomorphism*

$$HC^*_{pol}(\mathbb{C}\Gamma; \langle x\rangle) \cong HC^*(\mathbb{C}\Gamma; \langle x\rangle).$$

PROOF. Consider the commutative diagram

$$\begin{array}{ccccccc}
\to HH^n_{pol}(\mathbb{C}\Gamma; \langle x\rangle) & \xrightarrow{B} & HC^{n-1}_{pol}(\mathbb{C}\Gamma; \langle x\rangle) & \xrightarrow{S} & HC^{n+1}_{pol}(\mathbb{C}\Gamma; \langle x\rangle) & \longrightarrow & \cdots \\
\downarrow & & \downarrow & & \downarrow & & \\
\to HH^n(\mathbb{C}\Gamma; \langle x\rangle) & \xrightarrow{B} & HC^{n-1}(\mathbb{C}\Gamma; \langle x\rangle) & \xrightarrow{S} & HC^{n+1}(\mathbb{C}\Gamma; \langle x\rangle) & \longrightarrow & \cdots
\end{array}$$

By induction, starting with the trivial cases $n = -2$ and $n = -1$, the five lemma and Proposition 7.7 for HH imply the isomorphism for HC^n for all $n \geq -2$. □

Finally, for the pairing of cyclic cohomology with operator K-theory, we note that this factors through *periodic cyclic cohomology* HP^* (compare [**19**, Chapter 3, Section 3]). Recall that the (two periodic) $HP^*(\mathbb{C}\Gamma)$ is the direct limit

$$HP^k(\mathbb{C}\Gamma) = \text{dirlim}(HC^k(\mathbb{C}\Gamma) \xrightarrow{S} HC^{k+2}(\mathbb{C}\Gamma) \xrightarrow{S} \cdots) = HC^{k+2*}(\mathbb{C}\Gamma) \otimes_{\mathbb{C}[z]} \mathbb{C}$$

In the tensor product, the polynomial variable z of degree 2 acts on $HC^*(\mathbb{C}\Gamma)$ as the operator S and on $\mathbb{C}$ as 1.

The above considerations show that we also have a decomposition of $HP^*(\mathbb{C}\Gamma)$ with one term $HP^*(\mathbb{C}\Gamma; \langle x\rangle)$ for each conjugacy class $\langle x\rangle \subset \Gamma$ (compare [**19**, Chapter 3, Section 1]). Moreover, we also have polynomial growth versions of these periodic cohomology groups.

COROLLARY 7.9. *Let the group Γ be hyperbolic or of polynomial growth, $x \in \Gamma$. The inclusion induces an isomorphism*

$$HP^*_{pol}(\mathbb{C}\Gamma; \langle x\rangle) \cong HP^*(\mathbb{C}\Gamma; \langle x\rangle).$$

If x has finite order, then we have a functorial (for group homomorphisms) isomorphisms

$$HP^*_{pol}(\mathbb{C}\Gamma; \langle x\rangle) \cong \bigoplus_{k\subset\mathbb{Z}} H^{*+2k}(\Gamma_x; \mathbb{C}).$$

If, on the other hand, x has infinite order and Γ is hyperbolic then

$$HC^*_{pol}(\mathbb{C}\Gamma; \langle x\rangle) \cong \begin{cases} \mathbb{C}; & * = 0 \\ 0; & * > 0 \end{cases}; \qquad HP^*_{pol}(\mathbb{C}\Gamma; \langle x\rangle) = 0.$$

If x has infinite order and Γ has polynomial growth then at least $HP^*_{pol}(\mathbb{C}\Gamma; \langle x\rangle) = 0$.

PROOF. This is a direct consequence of Theorem 7.8 together with Theorem 7.1. From this last result we have, for $x \in \Gamma$ of finite order, that $HC^*(\mathbb{C}\Gamma; \langle x\rangle) = H^*(N_x; \mathbb{C}) \otimes \mathbb{C}[z]$, where $N_x = \Gamma_x/x^{\mathbb{Z}}$ and $x^{\mathbb{Z}}$ is the cyclic subgroup generated by x. However, the Leray-Serre spectral sequence for cohomology with complex coefficients of the short exact sequence $1 \to x^{\mathbb{Z}} \to \Gamma_x \to N_x \to 1$ degenerates on the E_2 page, as

$$H^*(x^{\mathbb{Z}}; \mathbb{C}) = \begin{cases} \mathbb{C}; & * = 0 \\ 0; & * \neq 0 \end{cases}$$

for the finite cyclic group $x^{\mathbb{Z}}$ and therefore the projection map induces an isomorphism $H^*(N_x; \mathbb{C}) \xrightarrow{\cong} H^*(\Gamma_x; \mathbb{C})$ which we are using implicitly. The action of S is given by multiplication by z which implies also the result for $HP^*(\mathbb{C}\Gamma; \langle x\rangle)$.

On the other hand, for $x \in \Gamma$ of infinite order $HC^*(\mathbb{C}\Gamma; \langle x\rangle) \cong H^*(N_x; \mathbb{C})$ where S acts raising degree by 2. If Γ is hyperbolic we know that Γ_x is a *finite* extension of the cyclic subgroup generated by x, therefore N_x is finite, compare [**27**, 8.5.M (1)]. It follows that its cohomology vanishes above degree 0 and the operator S of degree 2 is necessarily 0, so that the periodic cohomology vanishes.

If Γ has polynomial growth or equivalently is virtually nilpotent, also N_x is virtually nilpotent and in particular its rational cohomological dimension is finite. Consequently, $S^N = 0$ for N sufficiently large and its direct limit $HP^*(\mathbb{C}\Gamma; \langle x\rangle) = 0$. □

REMARK 7.10. Let us point out one subtlety when looking at the polynomially growing subcomplexes in this section. The results we have established are valid for one conjugacy class or for a finite number of them. On the other hand, the *full* cyclic cohomology groups are the unrestricted direct product of the contributions from the different conjugacy classes; consequently the above results have not been established for the full cyclic (co)homology groups with a global polynomial growth condition. However, relevant for us is the *periodic* cyclic cohomology, as the pairing with K-theory factors through the latter. If Γ is hyperbolic or Γ has polynomial growth, then Γ has only finitely many conjugacy classes of finite order (compare e.g. [**55**] and [**58**, Corollary 8]). Only those conjugacy classes contribute to $HP^*(\mathbb{C}\Gamma)$. For finite direct sums, polynomial growth conditions for each summand are equivalent to a global polynomial growth condition and we get the isomorphism $HP^*_{pol}(\mathbb{C}\Gamma) \xrightarrow{\cong} HP^*(\mathbb{C}\Gamma)$.

7.3. Extending cocycles for hyperbolic groups

Let Γ be a Gromov hyperbolic group. We want to use cyclic cohomology of Γ to obtain numerical primary and secondary invariants of Dirac operators. As explained in Section 7.1, for this we have to consider a suitable completion $\mathcal{A}\Gamma$

of $\mathbb{C}\Gamma$ which allows to construct the Chern character of Chapter 6, and then to argue that all cochains in a subcomplex of the cyclic cochain complex of $\mathbb{C}\Gamma$ extend to $\mathcal{A}\Gamma$. This is possible with a dense and holomorphically closed subalgebra of $C^*_{red}(\Gamma)$, that we will denote by $\mathcal{A}\Gamma$, constructed by Puschnigg in [**68**]. For this algebra we thus have that $K_*(\mathcal{A}\Gamma) \cong K_*(C^*_{red}(\Gamma))$. Puschnigg proves that any delocalized trace τ on $\mathbb{C}\Gamma$ which is supported on a finite number of conjugacy classes extends to $\mathcal{A}\Gamma$. Note that these are just certain cyclic cocycles of degree 0. We will show that the corresponding statement is true also for higher delocalized cochains, thus implementing the program explained in 7.1. Recall from Theorem 7.8 that the summand $HC^*(\mathbb{C}\Gamma;\langle x\rangle)$ supported on $\langle x\rangle$ is computed by $CC^\bullet_{pol}(\mathbb{C}\Gamma;\langle x\rangle)$, the subcomplex of cyclic cochains of polynomial growth supported on $CC_*(\mathbb{C}\Gamma;\langle x\rangle)$.

DEFINITION 7.11. If $\mathbb{C}\Gamma$ is densely included in a Fréchet algebra $\mathcal{A}\Gamma$, then we have an inclusion of $CC_\bullet(\mathbb{C}\Gamma;\langle x\rangle)$ into $CC_\bullet(\mathcal{A}\Gamma)$. We now define $CC_\bullet(\mathcal{A}\Gamma;\langle x\rangle)$ as the closure of $CC_\bullet(\mathbb{C}\Gamma;\langle x\rangle)$ inside $CC_\bullet(\mathcal{A}\Gamma)$, and correspondingly for $CC^\bullet$.

PROPOSITION 7.12. *Let Γ be a hyperbolic group, $x\in\Gamma$. The cochains of the subcomplex $CC^\bullet_{pol}(\mathbb{C}\Gamma;\langle x\rangle)$ extend continuously to $\mathcal{A}\Gamma$, giving rise to the canonical map*

$$HC^*(\mathbb{C}\Gamma;\langle x\rangle)\to HC^*(\mathcal{A}\Gamma)$$

which again is an inclusion as direct summand.

PROOF. We start with the duality pairing

$$c_{alg}\colon CC_\bullet(\mathbb{C}\Gamma)\otimes CC^\bullet_{pol}(\mathbb{C}\Gamma)\to\mathbb{C}$$

restricted to the cochains supported on $\langle x\rangle$,

$$c_{alg}\colon CC_\bullet(\mathbb{C}\Gamma)\otimes CC^\bullet_{pol}(\mathbb{C}\Gamma;\langle x\rangle)\to\mathbb{C}.$$

The assertion about the extension to $\mathcal{A}\Gamma$ means that we just have to show that c_{alg} extends continuously to $CC_\bullet(\mathcal{A}\Gamma)\otimes CC^\bullet_{pol}(\mathbb{C}\Gamma;\langle x\rangle)$. For this, we observe that the cochains supported on $\langle x\rangle$ pair non-trivially only with the chains supported on $\langle x\rangle$, i.e. c_{alg} factorizes over the projection $\pi_{\langle x\rangle}\colon CC_\bullet(\mathbb{C}\Gamma)\to CC_\bullet(\mathbb{C}\Gamma;\langle x\rangle)$. Next, by the very definition of rapid decay and of polynomial growth, we have a first extension

$$c_{RD}\colon CC_\bullet(l^1_{RD}\Gamma;\langle x\rangle)\otimes CC^\bullet_{pol}(\mathbb{C}\Gamma;\langle x\rangle)\to\mathbb{C}$$

of the algebraic pairing from $\mathbb{C}\Gamma$ to the rapid decay algebra $l^1_{RD}\Gamma$.

Next, when localized at $\langle x\rangle$, the cyclic chains of the l^1 rapid decay algebra $l^1_{RD}\Gamma$ and of $\mathcal{A}\Gamma$ coincide. More precisely, Puschnigg constructs in [**68**, Proposition 5.6, b)] a chain map $\iota_*\colon CC_\bullet(l^1_{RD}\Gamma;\langle x\rangle)\xrightarrow{\cong} CC_\bullet(\mathcal{A}\Gamma;\langle x\rangle)$ and shows that it is an isomorphism. Consequently, we get the pairing

$$c_{RD}\circ(\iota_*^{-1}\otimes\mathrm{id})\colon CC_\bullet(\mathcal{A}\Gamma;\langle x\rangle)\otimes CC^\bullet_{pol}(\mathbb{C}\Gamma;\langle x\rangle)\to\mathbb{C}.$$

Finally and crucially, Puschnigg constructs in the proof of [**68**, Proposition 5.6, a)] a (continuous) projection $\pi^{\mathcal{A}}_{\langle x\rangle}\colon CC_\bullet(\mathcal{A}\Gamma)\to CC_\bullet(\mathcal{A}\Gamma;\langle x\rangle)$, extending the corresponding projection $\pi_{\langle x\rangle}\colon CC_\bullet(\mathbb{C}\Gamma)\to CC_\bullet(\mathbb{C}\Gamma;\langle x\rangle)$. Note that he shows that this is well defined even for the entire (periodic) cyclic bicomplex. The argument works even more directly for the bicomplex computing usual cyclic homology, where we degree-wise only have to deal with finite sums.

The composition $c_{\mathcal{A}\Gamma} := c_{RD} \circ (\iota_*^{-1} \circ \pi_{\langle x\rangle}) \otimes \mathrm{id}\colon CC_\bullet(\mathcal{A}\Gamma) \otimes CC^\bullet_{pol}(\mathbb{C}\Gamma; \langle x\rangle) \to \mathbb{C}$ is then the desired extension of c_{alg}. The construction is summarized in the following commutative diagram:

(7.7)

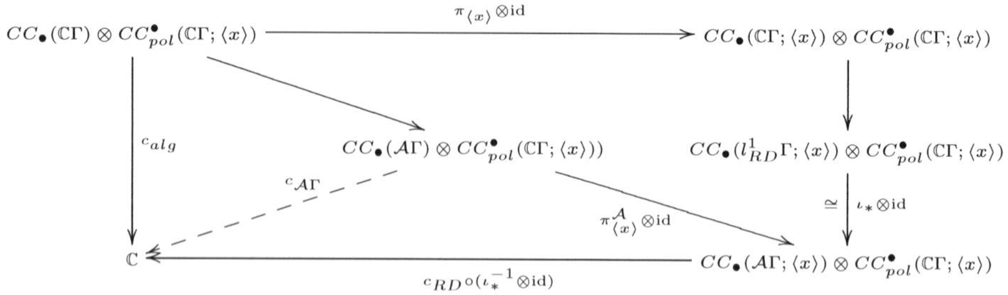

Using the identification $HC_{pol}(\mathbb{C}\Gamma; \langle x\rangle) \xrightarrow{\cong} HC^*(\mathbb{C}\Gamma; \langle x\rangle)$, $c_{\mathcal{A}\Gamma}$ corresponds dually to a homomorphism $HC^*(\mathbb{C}\Gamma; \langle x\rangle) \to HC^*(\mathcal{A}\Gamma)$ and the commutativity of the left triangle in (7.7) means that the composition

$$HC^*(\mathbb{C}\Gamma; \langle x\rangle) \to HC^*(\mathcal{A}\Gamma) \to HC^*(\mathbb{C}\Gamma)$$

is the split inclusion $HC^*(\mathbb{C}\Gamma; \langle x\rangle) \to HC^*(\mathbb{C}\Gamma)$ and therefore the first map

$$HC^*(\mathbb{C}\Gamma; \langle x\rangle) \to HC^*(\mathcal{A}\Gamma)$$

is split injective, too. □

Let us remark as an addendum the following propositions, the first of which follows immediately from the existence of the projection $\pi_{\langle x\rangle}$ of [**68**, Proposition 5.6].

PROPOSITION 7.13. *Let Γ be a hyperbolic group, $\mathcal{A}\Gamma$ the Puschnigg completion of $\mathbb{C}\Gamma$. For each $x \in \Gamma$, we have a split*

$$HC_*(\mathcal{A}\Gamma) \xrightarrow{(\pi_{\langle x\rangle})_*} HC_*(\mathcal{A}\Gamma; \langle x\rangle) \xleftarrow{\cong} HC_*(l^1_{RD}\Gamma; \langle x\rangle)$$

of the map $HC_*(\mathcal{A}\Gamma; \langle x\rangle) \to HC_*(\mathcal{A}\Gamma)$ induced by the (defining) inclusion of chain complexes.

PROPOSITION 7.14. *Let Γ be a hyperbolic group. The maps of Proposition* 7.12 *for the finitely many conjugacy classes of elements of finite order combine to give a canonical map*

$$HP^*(\mathbb{C}\Gamma) \to HP^*(\mathcal{A}\Gamma)$$

which is again an inclusion as direct summand.

PROOF. The argument of the proof of Proposition 7.12 works equally well if we replace $CC^\bullet_{pol}(\mathbb{C}; \langle x\rangle)$ by a direct sum of finitely many terms where the x represent pairwise different conjugacy classes. As Γ is hyperbolic, there are only finitely many conjugacy classes of finite order [**58**, Corollary 8]. We obtain the split injection $\bigoplus_{[x],x \text{ of finite order}} HC^*(\mathbb{C}\Gamma; \langle x\rangle) \to HC(\mathcal{A}\Gamma)$. By naturality, this split injection is compatible with the periodicity map S. Passing to the limit, we obtain the desired split injection

$$HP^*(\mathbb{C}\Gamma) = \bigoplus_{[x],x \text{ of finite order}} HP^*(\mathbb{C}\Gamma; \langle x\rangle) \to HP^*(\mathcal{A}\Gamma). \qquad \square$$

7.4. Extending cocycles for groups of polynomial growth

Let Γ by a group of polynomial growth. By [**38**, 3.1.8] these groups satisfy property (RD). By [**59**], it holds that $H^*_{pol}(\Gamma_x) \xrightarrow{\cong} H^*(\Gamma_x)$ (as each subgroup Γ_x of a group of polynomial growth has itself polynomial growth).

Consequently, $HP^*_{pol}(\mathbb{C}\Gamma) \to HP^*(\mathbb{C}\Gamma)$ is an isomorphism as this holds for each of the finitely many summands $H^*(\Gamma_x)$.

Let us choose for $\mathcal{A}\Gamma$ the Fréchet algebra $\mathcal{B}$ defined by Connes and Moscovici in [**20**, Section 6], which is dense and holomorphically closed in $C^*_{red}\Gamma$. Note that for groups of polynomial growth, $\mathcal{A}\Gamma$ coincides with the (l^2)-rapid decay algebra, compare [**38**, 3.1.7] and also with the l^1-rapid decay algebra $l^1_{RD}\Gamma$, compare [**39**, Remark 2.6 (2)]. Connes and Moscovici extend to $\mathcal{A}\Gamma$ the pairing with $H^*(\Gamma)$, the summand of $HP^*(\mathbb{C}\Gamma)$ corresponding to $x = e$ (the localized part of cyclic cohomology). The proof extends immediately to the full $HP^*(\mathbb{C}\Gamma)$, using that it works for each of the finitely many summands $H^*(\Gamma_x) = H^*_{pol}(\Gamma_x)$.

Equivalently, we have the following proposition.

PROPOSITION 7.15. *Let Γ be a group of polynomial growth. Let $\mathcal{A}\Gamma$ be the Connes-Moscovici algebra, a dense and holomorphically closed subalgebra of $C^*_{red}\Gamma$ containing $\mathbb{C}\Gamma$. Then we have a split injection*

$$HP^*(\mathbb{C}\Gamma) \to HP^*(\mathcal{A}\Gamma),$$

split by the restriction homomorphism.

We also have the splitting of the cyclic complex according to conjugacy classes, extending the one of the group ring, parallel to Proposition 7.13.

PROPOSITION 7.16. *Let Γ be a group of polynomial growth. Let $\mathcal{A}\Gamma$ be the Connes-Moscovici algebra, equal to $l^1_{RD}\Gamma$. For each $x \in \Gamma$, we have a split*

$$HC_*(\mathcal{A}\Gamma) \xrightarrow{(\pi_{\langle x\rangle})_*} HC_*(\mathcal{A}\Gamma; \langle x\rangle)$$

of the map $HC_*(\mathcal{A}\Gamma; \langle x\rangle) \to HC_*(\mathcal{A}\Gamma)$ induced by the (defining) inclusion of chain complexes.

PROOF. This is a well known fact, proven in [**36**, Corollary 1.4.2], using that for groups of polynomial growth the subalgebra $\mathcal{A}\Gamma$ relevant to us equals the l^1-rapid decay algebra treated there. □

7.5. Delocalized cyclic group homology and non-commutative de Rham homology

The considerations of Chapter 7.2 show, in particular, that the cyclic chain complex of a discrete group Γ has a "localized summand", namely $CC^e_*(\mathbb{C}\Gamma) := CC_*(\mathbb{C}\Gamma; \langle e\rangle)$ with complement the delocalized part $CC^{del}_*(\mathbb{C}\Gamma)$ and short (split) exact sequence

$$0 \to CC^e_\bullet(\mathbb{C}\Gamma) \to CC_\bullet(\mathbb{C}\Gamma) \to CC^{del}_\bullet(\mathbb{C}\Gamma) \to 0,$$

giving rise to a corresponding short split exact sequence of cyclic homology

$$0 \to HC^e_*(\mathbb{C}\Gamma) \to HC_*(\mathbb{C}\Gamma) \to HC^{del}_*(\mathbb{C}\Gamma) \to 0.$$

Assume now that $\mathcal{A}\Gamma$ is a Frèchet completion of $\mathbb{C}\Gamma$. Let $CC^e_*(\mathcal{A}\Gamma)$ denote the subcomplex $CC_*(\mathcal{A}\Gamma; \langle e\rangle)$, defined as in Section 7.3, then we define the delocalized cyclic complex via the short exact sequence

$$0 \to CC^e_\bullet(\mathcal{A}\Gamma) \to CC_\bullet(\mathcal{A}\Gamma) \to CC^{del}_\bullet(\mathcal{A}\Gamma) \to 0. \tag{7.8}$$

It is not guaranteed that this is split exact, but in any case we get an associated long exact sequence of localized and delocalized cyclic homology

$$\cdots \to HC^{del}_{*-1}(\mathcal{A}\Gamma) \to HC^e_*(\mathcal{A}\Gamma) \to HC_*(\mathcal{A}\Gamma) \to HC^{del}_*(\mathcal{A}\Gamma) \to \cdots$$

PROPOSITION 7.17. *If Γ is hyperbolic or of polynomial growth, then the long exact sequence for (de)localized cyclic homology splits into short exact sequence with a canonical split*

$$0 \longrightarrow HC^e_*(\mathcal{A}\Gamma) \longrightarrow HC_*(\mathcal{A}\Gamma) \longrightarrow HC^{del}_*(\mathcal{A}\Gamma) \longrightarrow 0\,.$$

with the splitting map $(\pi_{\langle e\rangle})_*\colon HC_*(\mathcal{A}\Gamma) \to HC^e_*(\mathcal{A}\Gamma)$ drawn as a curved arrow.

The corresponding result holds for cyclic cohomology.

PROOF. This is just the case $x = e$ of Propositions 7.13 and 7.16. □

Reduced cyclic homology. As mentioned in the introduction of this section, we want to relate the general results of Chapters 4, 5, and 6 to the computations of cyclic (co)homology established in the current section. It is well known that this is possible: non-commutative de Rham homology canonically embeds in cyclic homology.

Instead of deriving a full account of this, which also works for the relative homology we emphasize in this paper, we just want to cite the relevant results in the literature [**40**, Théorème 2.15]. Unfortunately, there is one detail to take into account: the embedding works with *reduced* cyclic homology.[1]

We omit here the explicit definition [**40**, Section 2.9] and instead observe that for the cases relevant to us the reduced homology $\overline{HC}_*(\mathcal{A}\Gamma)$ is easily described:

PROPOSITION 7.18. *Let the group Γ be hyperbolic or of polynomial growth. As above, let $\mathcal{A}\Gamma$ be the Puschnigg algebra if Γ is hyperbolic or the rapid decay algebra if Γ has polynomial growth.*

Then we have a canonical decomposition

$$HC_*(\mathcal{A}\Gamma) = HC_*(\mathbb{C}) \oplus \overline{HC}_*(\mathcal{A}\Gamma)$$

with projection map $\epsilon_*\colon HC_*(\mathcal{A}\Gamma) \to HC_*(\mathbb{C})$ splitting the map induced by the inclusion $\mathbb{C} \to \mathcal{A}\Gamma$.

$$HC_*(\mathcal{A}\Gamma) = \underbrace{\underbrace{HC^{del}_*(\mathcal{A}\Gamma)}_{=\overline{HC}^{del}_*(\mathcal{A}\Gamma)} \oplus \overline{HC}^e_*(\mathcal{A}\Gamma)}_{=\overline{HC}_*(\mathcal{A}\Gamma)} \oplus HC_*(\mathbb{C});$$

$$\overline{HC}^e_*(\mathcal{A}\Gamma) \oplus HC_*(\mathbb{C}) = HC^e_*(\mathcal{A}\Gamma)$$

[1] "Reduced" here refers to quotient by terms coming from the ground field $\mathbb{C}$, corresponding to a base point in topology. This must not be confused with notion of reduced homology when chain complexes of topological vector spaces are considered and homology is defined by taking the quotient by the *closure* of the image of the differential. We never use this second meaning in this paper.

PROOF. It suffices to construct the split ϵ_*. For this, we use the split $\pi_{\langle e\rangle}\colon HC_*(\mathcal{A}\Gamma) \to HC_*(\mathcal{A}\Gamma;\langle e\rangle) = HC_*(l^1_{RD}\Gamma;\langle e\rangle)$ and compose with the morphism induced by the canonical augmentation map $\epsilon\colon l^1\Gamma \to \mathbb{C}; \sum a_\gamma\gamma \mapsto \sum a_\gamma$, which is defined because $l^1_{RD}\Gamma \subset l^1\Gamma$. □

COROLLARY 7.19. *Let the group Γ be hyperbolic or of polynomial growth and $\mathcal{A}\Gamma$ the smooth subalgebras of $C^*_{red}\Gamma$ as before. We get a commutative diagram, mapping the localized/delocalized non-commutative de Rham homology sequence to the localized/delocalized cyclic homology sequence*

(7.9)
$$\begin{array}{ccccccccc}
\longrightarrow & H^e_*(\mathcal{A}\Gamma) & \longrightarrow & H_*(\mathcal{A}\Gamma) & \longrightarrow & H^{del}_*(\mathcal{A}\Gamma) & \longrightarrow & H^e_{*-1}(\mathcal{A}\Gamma) & \\
 & \downarrow & & \downarrow & & \downarrow & & & \\
0 \longrightarrow & \overline{HC}^e_*(\mathcal{A}\Gamma) & \longrightarrow & \overline{HC}_*(\mathcal{A}\Gamma) & \longrightarrow & HC^{del}_*(\mathcal{A}\Gamma) & \longrightarrow & 0 &
\end{array}$$

Note that we have established that the sequence of cyclic homology is split short exact in our situation in Propositions 7.13 and 7.16. We suspect that the same is true for the non-commutative de Rham sequence, but we don't make use of this and therefore don't discuss this issue here. Note, moreover, that such a splitting would imply immediately that all three arrows are injective (after passage to reduced de Rham homology), because the middle one is. Again, we don't use this injectivity and therefore don't discuss it further.

PROOF. This follows from the proof of the inclusion $\overline{H}_*(\mathcal{A}\Gamma) \hookrightarrow \overline{HC}_*(\mathcal{A}\Gamma)$ of [**50**, Section 2.6.4] or [**40**, Théorème 2.15] (where we note that $\overline{H}_*(\mathcal{A})$ simply is the quotient of $H_*(\mathcal{A})$ by the image of $H_*(\mathbb{C})$, where the latter is concentrated in degree 0 and equal to $\mathbb{C}$ there. In particular, $\overline{H}_n(\mathcal{A}\Gamma) = H_n(\mathcal{A}\Gamma)$ for $n \geq 1$).

The inclusion ultimately is obtained by mapping a form $a_0da_1\dots da_n$ to $a_0 \otimes \dots \otimes a_n$. It follows immediately that $H^e_*(\mathcal{A}\Gamma)$ is mapped to $\overline{HC}^e_*(\mathcal{A}\Gamma)$, and the left vertical arrow of (7.9) is defined. Moreover, inspecting the proof then shows that also the right vertical map is well defined and of course makes the diagram commutative. □

REMARK 7.20. This fact is key in what follows, since it allows to see elements in the image of Ch^{del}_Γ, that is elements in $H^{del}_*(\mathcal{A}\Gamma)$, as elements in the cyclic homology of $\mathcal{A}\Gamma$ and then to pair them with delocalized cyclic cocycles.

Of course, the fact that we can pair with cyclic cocycles is in the end well known, but we preferred to give a formally correct derivation of this via the Chern character map, which in our approach a priori takes values in non-commutative de Rham homology.

REMARK 7.21. We believe that the treatment of Corollary 7.19 could be systematized and improved in the following two ways:

- It is a bit unfortunate that the target of the maps in Corollary 7.19 is *reduced* cyclic homology, because we know that we can pair K-theory also with unreduced cohomology. This is due to the classical definition of de Rham homology defining $d1 = 0$, which forces to implement a similar relation in the target cyclic homology groups, leading to the occurrence of $\overline{HC}$.

  A logical remedy, very much in the spirit of the definition of K-theory, would be to work with a version of non-commutative de Rham homology which first augments the algebra and then passes to the reduced homology of this augmented algebra. This works uniformly for unital and non-unital algebras, is completely in line with the definition of K-theory, and avoids the appearance of reduced cyclic homology in the comparison map.

- In the derivation of Corollary 7.19, we tried to be as brief as possible, reducing directly to the known result for $H_*(\mathcal{A}\Gamma)$. Along the way, we used the splitting of Proposition 7.18. We believe that a more systematic and general direct treatment could be interesting, which should work without further splitting assumptions.

7.6. Higher rho numbers

In this subsection, let the group Γ be hyperbolic or of polynomial growth. Using Corollary 7.9, Proposition 7.14, Proposition 7.15 and Remark 7.20 we obtain immediately the following result.

THEOREM 7.22. *Let the group Γ be hyperbolic or of polynomial growth and $e \neq x \in \Gamma$ and let $\widetilde{M}$ be a Galois covering of a smooth compact manifold M with covering group Γ. Then there exists a well-defined pairing*

$$\mathrm{S}^\Gamma_*(\widetilde{M}) \times HP^{*-1}(\mathbb{C}\Gamma; \langle x \rangle) = K_* \left(C(M) \to \Psi^0_\Gamma(\widetilde{M}) \right) \times HP^{*-1}(\mathbb{C}\Gamma; \langle x \rangle) \to \mathbb{C}$$

given by

$$(\xi, [\tau]) \mapsto \langle \mathrm{Ch}^{del}_\Gamma(\xi), \tau \rangle$$

for $\xi \in K_{*+1} \left(C(M) \to \Psi^0_\Gamma(\widetilde{M}) \right)$ and $\tau \in CC^{[*-1]}_{pol}(\mathbb{C}\Gamma; \langle x \rangle)$.

REMARK 7.23. This pairing is compatible with the usual pairing between $K_{*-1}(C^*_{red}\Gamma)$ and cyclic cohomology: for $\xi \in K_{*+1}(C^*_{red}\Gamma)$ and $[\tau] \in HC^{[*-1]}(\mathbb{C}\Gamma; \langle x \rangle)$ we have $\langle \xi, [\tau] \rangle = \langle s(\xi), [\tau] \rangle$, where $s \colon K_{*-1}(C^*_{red}\Gamma) \to \mathrm{S}^\Gamma_*(\widetilde{M})$ is as in (6.24).

DEFINITION 7.24. Let $\widetilde{D}$ be a generalized Dirac operator which is Γ-equivariant on the Galois Γ-covering $\widetilde{M}$ of a compact smooth manifold M of dimension n without boundary. Suppose that $\widetilde{D}$ is L^2-invertible. Then define its higher rho number associated to $[\tau] \in HC^{[n-1]}(\mathbb{C}\Gamma; \langle x \rangle)$ as

$$\varrho_\tau(\widetilde{D}) := \left\langle \mathrm{Ch}^{del}_\Gamma \left(\varrho(\widetilde{D}) \right), \tau \right\rangle \in \mathbb{C},$$

where $\varrho(\widetilde{D})$ is defined as in Definition 3.4.

DEFINITION 7.25. Let g be a metric with positive scalar curvature on a spin manifold M of dimension n with fundamental group Γ. Define the higher rho number associated to $[\tau] \in HC^{[n-1]}(\mathbb{C}\Gamma; \langle x \rangle)$ as

$$\varrho_\tau(g) := \langle \mathrm{Ch}^{del}_\Gamma \left(\varrho(g) \right), \tau \rangle \in \mathbb{C}$$

where $\varrho(g) \in K_n \left(C(M) \to \Psi^0_\Gamma(\widetilde{M}) \right)$ is the rho class of the spin Dirac operator for the metric g.

Recall from (2.4) that, if the dimension of M is odd, we have a surjective map

$$K_0(\Psi^0_{\mathcal{A}\Gamma}(\widetilde{M})) \to K_1(C(M) \to \Psi^0_{\mathcal{A}\Gamma}(\widetilde{M})),$$

given by the composition of the suspension isomorphism and the natural inclusion. Moreover, by Lemma 4.22 the triangle

$$\begin{array}{ccc} K_0(\Psi^0_{\mathcal{A}\Gamma}(\widetilde{M})) & \longrightarrow & K_1(C(M) \to \Psi^0_{\mathcal{A}\Gamma}(\widetilde{M})) \\ & \searrow^{\overline{\mathrm{TR}}^{del} \circ \mathrm{Ch}} & \downarrow \mathrm{Ch}^{del}_{\Gamma} \\ & & H^{del}_{\mathrm{even}}(\mathcal{A}\Gamma) \end{array}$$

is commutative. Therefore, given a metric with positive scalar curvature g on an odd dimensional manifold M, we can express $\mathrm{Ch}^{del}_{\Gamma}(\varrho(g))$ by $\overline{\mathrm{TR}}^{del}(\mathrm{Ch}[\pi_{\geq}])$, as in Section 3.2. In order to calculate the higher rho number associated to an even dimensional cyclic cocycle τ, we then proceed as in the following proposition.

PROPOSITION 7.26. *Let τ be a cyclic cocycle of degree $2k$ of polynomial growth with support contained in a conjugacy class $\langle x\rangle \neq \langle e\rangle$ of Γ. Then the explicit integral formula is given by the following expression*

$$\varrho_\tau(g) = \sum_{g_0\cdots g_{2k}\in\langle x\rangle} \frac{1}{k!}\int_{\widetilde{M}\times\cdots\times\widetilde{M}} \mathrm{Tr}\,(\chi(x_0)P(x_0g_0,x_1)h(x_1)P(x_1g_1,x_2)\cdots \\ \cdots h(x_{2k})P(x_{2k}g_{2k},x_0))\ \tau(g_0,\dots,g_{2k})\, dvol_g^{2k+1} \tag{7.10}$$

where P is the Schwartz kernel of $\pi_{\geq}(\widetilde{D}_g)$ and where χ is the characteristic function of a fundamental domain $\mathcal{F}$ for the action of Γ on $\widetilde{M}$, and where h is the cutoff function of (6.3).

PROOF. First observe, using the notation of Chapter 6, as P is a projector and d a derivation, that we have $dP = d(P^2) = Pd + (dP)P$, or $(dP)P = dP - PdP$. Applying this twice we get $PdPdPdPdP = (PdPdP)^2$. Moreover, by the definition of d we have $dP = [\nabla^{Lott}, P] + PX + XP$ with the auxiliary formal variable X with $X^2 = \Theta$ and $PXP = 0$. Substituting, we obtain

$$PdPdP = P[\nabla^{Lott}, P][\nabla^{Lott}, P] + P\Theta P.$$

Moreover, using that $P[\nabla^{Lott}, P]P = P \circ \nabla^{Lott} \circ P - P \circ \nabla^{Lott} \circ P = 0$,

$$\begin{aligned} P[\nabla^{Lott}, P][\nabla^{Lott}, P] + P\Theta P &= P[\nabla^{Lott}, P][\nabla^{Lott}, P] + P(\nabla^{Lott})^2 P \\ &\quad + P[\nabla^{Lott}, P]P\nabla^{Lott} \\ &= P[\nabla^{Lott}, P]\nabla^{Lott}P + P(\nabla^{Lott})^2 P \\ &= P\nabla^{Lott}P\nabla^{Lott}P \\ &= (P\nabla^{Lott}P)^2. \end{aligned} \tag{7.11}$$

Hence we have that $\mathrm{Ch}^{del}_{\Gamma,k}(\pi_{\geq})$ is given by

$$\overline{\mathrm{TR}}^{del}((P\nabla^{Lott}P)^{2k}).$$

In order to obtain an explicit formula for this we use the version of ∇^{Lott} given by (6.5). A direct calculation then gives (7.10). □

The corresponding result applies to any L^2-invertible Γ-equivariant Dirac type operator $\widetilde{D}$ on an odd dimensional manifold $\widetilde{M}$.

EXAMPLE 7.27. Let τ be a 0-cocycle supported on the conjugacy class $\langle x \rangle$ in Γ. Then

$$\varrho_\tau(\widetilde{D}) = \sum_{\gamma \in \langle x \rangle} \int_{\mathcal{F}} \mathrm{Tr}(\pi_{\geq}(\widetilde{D})(x\gamma, x)) d\mathrm{vol}(x)\tau(\gamma)$$

which is, up to a normalization factor, exactly the delocalized η-invariant of Lott obtained in [**52**, Section 4.10.1]. This example also appears in [**18**].

We have a corresponding result for the numeric rho class of a positive scalar curvature metric on an even dimensional spin manifold associated to an odd degree cyclic cohomology class $[\tau]$, which is not quite as explicit and which is worked out in Proposition 7.31.

EXAMPLE 7.28. Let us now consider the case of an even dimensional spin Riemannian manifold (M, g) of positive scalar curvature with fundamental group Γ. We work out explicitly the degree 1 part of the Chern character of the rho class. Recall that the rho class of the metric g in $K_0(C(M) \to \Psi^0_{\mathcal{A}\Gamma}(\widetilde{M}))$ is given by the path of projections

$$p_t = \begin{pmatrix} \cos^2\left(\frac{\pi}{2}t\right) e_0 & \cos\cdot\sin\left(\frac{\pi}{2}t\right) u^* \\ \cos\cdot\sin\left(\frac{\pi}{2}t\right) u & \sin^2\left(\frac{\pi}{2}t\right) e_1 \end{pmatrix} \quad \text{on the module} \quad \Psi^0_{\mathcal{A}\Gamma}(\widetilde{M})^n \oplus \Psi^0_{\mathcal{A}\Gamma}(\widetilde{M})^n,$$

for some $n \in \mathbb{N}$ and $t \in [0,1]$. Here e_0 and e_1 are the projections over $C(M)$ associated to the spinor vector bundles S_+ and S_- respectively; whereas u and u^* are the partial isometries $u = e_1 \operatorname{sgn}(\widetilde{D})_+ e_0$ and consequently $u^* = e_0 \operatorname{sgn}(\widetilde{D})_- e_1$, with $\tilde{D}$ the (invertible) Dirac operator on $\widetilde{M}$. By Proposition 4.15 and Remark 4.16, the degree 1 part of $\mathrm{Ch}^{del}_\Gamma(\varrho(\widetilde{D}))$ is given by

$$\overline{\mathrm{TR}}^{del}\left(\int_0^1 \mathrm{Tr}\left((2p_t - 1)\dot{p}_t dp_t\right) dt\right). \tag{7.12}$$

Observe first that

$$\dot{p}_t = \begin{pmatrix} -\pi\cos\cdot\sin\left(\frac{\pi}{2}t\right) e_0 & \frac{\pi}{2}(\cos^2 - \sin^2)\left(\frac{\pi}{2}t\right) u^* \\ \frac{\pi}{2}(\cos^2 - \sin^2)\left(\frac{\pi}{2}t\right) u & \pi\cos\cdot\sin\left(\frac{\pi}{2}t\right) e_1 \end{pmatrix}.$$

Moreover recall by Definition 6.13 that the differential d involves also the auxiliary variable X and that $\overline{\mathrm{TR}}^{del}$ is zero on the terms in which X appears precisely once. Hence we have that dp_t is given by

$$\begin{pmatrix} 0 & \cos\cdot\sin\left(\frac{\pi}{2}t\right)[\nabla, u^*] \\ \cos\cdot\sin\left(\frac{\pi}{2}t\right)[\nabla, u] & 0 \end{pmatrix},$$

up to terms involving X.

Now an elementary but tedious calculation shows that the integrand of (7.12) is given by

$$\begin{aligned} &\pi\cos^3\cdot\sin(\cos^2 - \sin^2)\left(\frac{\pi}{2}t\right) e_0 u^*[\nabla, u] - \frac{\pi}{2}\cos\cdot\sin\cdot(\cos^2 - \sin^2)\left(\frac{\pi}{2}t\right) u^*[\nabla, u] \\ &\quad + \pi\cos^3\cdot\sin^3\left(\frac{\pi}{2}t\right) u^* e_1[\nabla, u] \\ &\quad - \pi\cos^3\cdot\sin^3\left(\frac{\pi}{2}t\right) u e_0[\nabla, u^*] + \pi\cos\cdot\sin^3\left(\frac{\pi}{2}t\right) e_1 u[\nabla, u^*] \\ &\quad - \frac{\pi}{2}\cos\cdot\sin^3\cdot(\cos^2 - \sin^2)\left(\frac{\pi}{2}t\right) u[\nabla, u^*]. \end{aligned}$$

Observe that the second and the last term have zero integral. By using that $u^*e_1 = e_0u^*$ and that $ue_0 = e_1u$ we obtain that the integrand in (7.12) is given by

$$\pi\cos^3\cdot\sin\left(\frac{\pi}{2}t\right)e_0u^*[\nabla,u] - \pi\cos\cdot\sin^3\left(\frac{\pi}{2}t\right)e_1u[\nabla,u^*]$$

which in turns leads to $\frac{1}{2}\mathrm{TR}^{del}(e_0u^*[\nabla,u] - e_1u[\nabla,u^*])$. Now observe that $e_0u^*[\nabla,u] = \mathrm{sgn}(\widetilde{D})_-[\nabla,\mathrm{sgn}(\widetilde{D})_+]$ and that $e_1u[\nabla,u^*] = \mathrm{sgn}(\widetilde{D})_+[\nabla,\mathrm{sgn}(\widetilde{D})_-]$. Thus we finally obtain that the degree 1 part of the delocalized Chern character of $\varrho(g)$ is given by

$$\begin{aligned}\mathrm{Ch}^{del}_\Gamma(\varrho(g))_{[1]} &= \frac{1}{2}\mathrm{STR}^{del}\left(\mathrm{sgn}(\widetilde{D})[\nabla,\mathrm{sgn}(\widetilde{D})]\right)\\ &= -\frac{1}{2}\mathrm{STR}^{del}\left([\nabla,\mathrm{sgn}(\widetilde{D})]\,\mathrm{sgn}(\widetilde{D})\right),\end{aligned}$$

where STR^{del} denotes the delocalized supertrace with respect to the natural grading of the spinor bundle of an even dimensional manifold and the minus in the last term appeared because $\mathrm{sgn}(\widetilde{D})$ is odd with respect to this grading. Using (6.8), we can explicitly write down the Schwartz kernel of $\mathrm{sgn}(\widetilde{D})[\nabla,\mathrm{sgn}(\widetilde{D})]$, obtaining

$$\begin{aligned}&\mathrm{STR}^{del}(\mathrm{sgn}(\tilde{D})[\nabla,\mathrm{sgn}(\tilde{D})])\\ &\quad= \sum_{\gamma_0,\gamma_1\neq e}\int_{\widetilde{M}\times\widetilde{M}}\mathrm{STr}(\chi(x)\,\mathrm{sgn}(\widetilde{D})(x\gamma_0,y)h(y)\,\mathrm{sgn}(\widetilde{D})(y\gamma_1,x))dxdy\;\gamma_0 d\gamma_1.\end{aligned}$$

with χ the characteristic function of a fundamental domain of the covering projection $\widetilde{M}\to M$.

COROLLARY 7.29. *In the setting of Example* 7.28, *let τ be a cyclic cocycle of degree* 1 *of polynomial growth with support the conjugacy class $\langle x\rangle\neq\langle e\rangle$ of Γ. Then*

$$\begin{aligned}\varrho_\tau(g) = \frac{1}{2}\sum_{\gamma_0\gamma_1\in\langle x\rangle}\int_{\widetilde{M}\times\tilde{M}}&\mathrm{STr}(\chi(x_0)\,\mathrm{sgn}(\tilde{D})(x_0\gamma_0,x_1)h(x_1)\,\mathrm{sgn}(\tilde{D})(x_1\gamma_1,x_0))\tau\\ &\cdot(\gamma_0,\gamma_1)\,dx_0\,dx_1.\end{aligned}$$

COROLLARY 7.30. *In the situation of Example* 7.28, *let τ be again a delocalized cyclic cocycle of degree* 1 *supported on a conjugacy class $\langle x\rangle\neq\langle e\rangle$. Then the delocalized higher rho invariant $\varrho_\tau(g)$ coincides with the one constructed in* [**52**, Section 4.10.2], *i.e.*

$$\varrho_\tau(g) = -\frac{1}{2}\langle\mathrm{STR}^{del}\left([\nabla,\widetilde{D}]\widetilde{D}^{-1}\right),\tau\rangle$$

PROOF. In the following, let $\sim$ denote equality up super-commutators. Then

$$
\begin{aligned}
&[\nabla, \operatorname{sgn}(\widetilde{D})]\operatorname{sgn}(\widetilde{D}) \\
&-[\nabla, \operatorname{sgn}(\widetilde{D})]\operatorname{sgn}(\widetilde{D}) + \frac{1}{2}[\nabla, |\widetilde{D}|]|\widetilde{D}^{-1}| - \frac{1}{2}[\nabla, |\widetilde{D}|]|\widetilde{D}^{-1}| \\
=&\frac{1}{2}\left(|\widetilde{D}^{-1}||\widetilde{D}|[\nabla, \operatorname{sgn}(\widetilde{D})]\operatorname{sgn}(\widetilde{D}) + [\nabla, |\widetilde{D}|]\operatorname{sgn}(\widetilde{D})\operatorname{sgn}(\widetilde{D})|\widetilde{D}^{-1}|\right) \\
&+\frac{1}{2}\left([\nabla, \operatorname{sgn}(\widetilde{D})]|\widetilde{D}||\widetilde{D}^{-1}|\operatorname{sgn}(\widetilde{D}) + [\nabla, |\widetilde{D}|]|\widetilde{D}^{-1}|\operatorname{sgn}(\widetilde{D})\operatorname{sgn}(\widetilde{D})\right) \\
\sim&\frac{1}{2}\left(|\widetilde{D}|[\nabla, \operatorname{sgn}(\widetilde{D})]\operatorname{sgn}(\widetilde{D})|\widetilde{D}^{-1}| + [\nabla, |\widetilde{D}|]\operatorname{sgn}(\widetilde{D})\operatorname{sgn}(\widetilde{D})|\widetilde{D}^{-1}|\right) \\
&+\frac{1}{2}\left([\nabla, \operatorname{sgn}(\widetilde{D})]|\widetilde{D}||\widetilde{D}^{-1}|\operatorname{sgn}(\widetilde{D}) + \operatorname{sgn}(\widetilde{D})[\nabla, |\widetilde{D}|]|\widetilde{D}^{-1}|\operatorname{sgn}(\widetilde{D})\right) \\
=&\frac{1}{2}[\nabla, |\widetilde{D}|\operatorname{sgn}(\widetilde{D})]\operatorname{sgn}(\widetilde{D})|\widetilde{D}^{-1}| + \frac{1}{2}[\nabla, \operatorname{sgn}(\widetilde{D})|\widetilde{D}|]|\widetilde{D}^{-1}|\operatorname{sgn}(\widetilde{D}) \\
=&[\nabla, \widetilde{D}]\widetilde{D}^{-1}.
\end{aligned}
\tag{7.13}
$$

where in the second to last line we used the Leibniz formula for commutators. Note that we apply super-commutators in the algebra of all pseudodifferential operators, not only operators of degree 0. However, because our *delocalized* supertrace has the trace property on this larger algebra by Proposition 6.16, Equation (7.13) implies that the degree 1 part of the delocalized Chern character of $\varrho(g)$ in the even dimensional case is given by

$$
\operatorname{Ch}^{del}_{\Gamma}(\varrho(g))_{[1]} = -\frac{1}{2}\operatorname{STR}^{del}\left([\nabla, \operatorname{sgn}(\tilde{D})]\operatorname{sgn}(\tilde{D})\right) = -\frac{1}{2}\operatorname{STR}^{del}\left([\nabla, \widetilde{D}]\widetilde{D}^{-1}\right)
$$

which, after paired with a cyclic cocycle of degree 1, gives exactly the expression appearing in [**52**, Section 4.10.2]. □

We now pass to the general case of cyclic cocycles of odd degree, say $2k+1$. First of all, we need to identify the $2k+1$-degree part of $\operatorname{Ch}^{del}_{\Gamma}(\varrho(\widetilde{D}))$; recall that this is given by

$$
\overline{\operatorname{TR}}^{del}\left(\int_0^1 \operatorname{Tr}\left((2p_t - 1)\dot{p}_t(dp_t)^{2k+1}\right) dt\right). \tag{7.14}
$$

Observe that

$$
dp_t = [\nabla, p_t] + Xp_t + p_tX
$$

and then that

$$
(dp_t)^2 = [\nabla, p_t]^2 + [\nabla, p_t]p_tX + Xp_tX + Xp_t[\nabla, p_t] + p_t\Theta p_t.
$$

Taking into account that $\mathbf{T}_1X\mathbf{T}_2 = 0$ and that $\overline{\operatorname{TR}}^{del}$ is zero on operators of the form $\mathbf{T}_1X$ or $X\mathbf{T}_2$, we can observe the following fact: $(2p_t - 1)\dot{p}_t dp_t = (2p_t - 1)\dot{p}_t([\nabla, p_t] + p_tX)$, hence, when we calculate $dp_t(dp_t)^{2k}$ inside the integrand of (7.14), we are reduced to consider

$$
([\nabla, p_t] + p_tX)([\nabla, p_t]^2 + [\nabla, p_t]p_tX + Xp_tX + Xp_t[\nabla, p_t] + p_t\Theta p_t)^{2k}.
$$

Thus, once we apply the delocalized trace $\overline{\operatorname{TR}}^{del}$ the only products that survive are the following ones:

- if they start with $[\nabla, p_t]$ then they contain after this factor, in all permutations, elements of the form $[\nabla, p_t]^{2j}$, $[\nabla, p_t]p_t X(Xp_tX)^h Xp_t[\nabla, p_t] = [\nabla, p_t](p_t\Theta p_t)^{h+1}[\nabla, p_t]$ or $(p_t\Theta p_t)^l$, so that the total degree is $2k+1$;
- if they start with p_tX, then, bearing in mind that $\mathbf{T}_1 X \mathbf{T}_2 = 0$, such a term needs to be multiplied by an element of the form $(Xp_tX)^h Xp_t[\nabla, p_t]$. We then obtain in the first place $(p_t\Theta p_t)^{h+1}[\nabla, p_t]$, which will be followed by element of the form as in the previous point, in all permutations, so that the total degree is $2k+1$.

Summarizing $(2p_t - 1)\dot{p}_t(dp_t)^{2k+1}$ is the sum of products starting with $(2p_t - 1)\dot{p}_t$, followed by all the permutations of $2h+1$ copies of $[\nabla, p_t]$ and l copies of $p_t\Theta p_t$, for all $h, l \in \mathbb{N}$ such that $2l + 2h + 1 = 2k + 1$. Let us denote by Q_t the Schwartz kernel of $(2p_t - 1)\dot{p}_t$ and by P_t the Schwartz kernel of p_t. Then we know that the Schwartz kernel of $[\nabla, p_t]$ is given by

$$\sum_{\gamma\in\Gamma}(h(x) - R^*_{\gamma^{-1}}h(y))R^*_{(\gamma,e)}P_t(x,y)d\gamma \tag{7.15}$$

whereas the Schwartz kernel of $p_t\Theta p_t$ is given by

$$\begin{aligned}&\sum_{\gamma_1,\gamma_2\in\Gamma}\int_{\widetilde{M}} P_t(x,z)h(z)R^*_{\gamma_1}h(z)R^*_{(\gamma_1\gamma_2,e)}P_t(z,y)dvol_g(z)\otimes d\gamma_1 d\gamma_2\\ &= \sum_{\gamma_1,\gamma_2\in\Gamma}\int_{\widetilde{M}} R^*_{(\gamma_1,e)}P_t(x,z)R^*_{\gamma_1^{-1}}h(z)h(z)R^*_{(e,\gamma_2^{-1})}P_t(z,y)dvol_g(z)\otimes d\gamma_1 d\gamma_2\end{aligned} \tag{7.16}$$

Without working out the precise result, we obtain the following abstract formula for (7.14).

PROPOSITION 7.31.

$$\begin{aligned}\mathrm{Ch}^{del}_\Gamma(\varrho(\widetilde{D}))_{2k+1} = &\sum_{\gamma_0\ldots\gamma_{2k+1}\neq e}\int_0^1 dt\int_{\widetilde{M}\times\cdots\times\widetilde{M}}\mathrm{Tr}\Big(\chi(x_0)R^*_{\gamma_0}Q_t(x_0,x_1)\\ &\sum R^*_{(\alpha^i_0,\ldots,\alpha^i_{2k+1})}C^i_t(x_1,\ldots,x_{2k+1},x_0)\Big)\, dvol_g^{2k+1}\gamma_0 d\gamma_1\ldots d\gamma_{2k+1}\end{aligned} \tag{7.17}$$

where α^i_j is a product of elements in $\{\gamma_0, \gamma_1, \ldots, \gamma_{2k+1}\}$ and the C^i_t are the convolution of all the permutation of $2h_i+1$ copies of (7.15) and l_i copies of (7.16), for all $h_i, l_i \in \mathbb{N}$ such that $l_i + h_i = k$ (the central sum is over all these terms).

Now it is immediate to obtain an expression like the one in formula (7.10) when we pair with τ, a cyclic cocycle of degree $2k+1$ of polynomial growth with support contained in a conjugacy class $\langle x\rangle$ of Γ.

CHAPTER 8

Higher rho numbers from relative cohomology $H^*(M \to B\Gamma)$

When looking at the Higson-Roe sequence $\cdots \to K_*(M) \to K_*(C^*\Gamma) \to \mathrm{S}^{\Gamma}_{*-1}(\widetilde{M}) \to K_{*-1}(M) \to \cdots$ we realize that, at least morally, $K(C^*\Gamma)$ is closely related to $K_*(B\Gamma)$. Consequently, there should be a second source of information on the relative term (the structure set $\mathrm{S}^{\Gamma}_{*-1}(\widetilde{M})$) in terms of the difference between M and $B\Gamma$. In this section, we make this idea precise: we construct a pairing between the analytic structure set $\mathrm{S}^{\Gamma}_{*}(\widetilde{M})$ and the relative cohomology $H^*(M \to B\Gamma; \mathbb{R})$, at least for suitable groups. To achieve this and to arrive at explicit formulae, we start with a convenient cocycle model for $H^*(M \to B\Gamma; \mathbb{R})$, using suitable Alexander-Spanier cochains on $\widetilde{M}$.

8.1. Alexander-Spanier cochains and singular cohomology

DEFINITION 8.1. Let $C^q_{AS}(\widetilde{M})^{\Gamma}$ be the vector space of all (not necessarily continuous) functions $\varphi\colon \widetilde{M}^{q+1} \to \mathbb{R}$ which are invariant with respect to the diagonal action of Γ on $\widetilde{M}^{q+1}$. Define $\delta\colon C^q_{AS}(\widetilde{M})^{\Gamma} \to C^{q+1}_{AS}(\widetilde{M})^{\Gamma}$ by the standard formula

$$(\delta\varphi)(x_0, \ldots, x_{q+1}) = \sum_{i=0}^{q+1} (-1)^i \varphi(x_0, \ldots, \hat{x}_i, \ldots, x_{q+1})$$

which makes $\{C^{\bullet}_{AS}(\widetilde{M})^{\Gamma}, \delta\}$ a cochain complex.

An important subcomplex of $C^{\bullet}_{AS}(\widetilde{M})^{\Gamma}$ is given by locally zero cochains. An element $\varphi \in C^{\bullet}_{AS}(\widetilde{M})^{\Gamma}$ is said to be locally zero if there is an open neighborhood of the diagonal $\widetilde{M} \subset \widetilde{M}^{q+1}$ on which φ vanishes. We denote by $C^q_{AS,0}(\widetilde{M})^{\Gamma}$ the vector space of all such functions, forming a subcomplex of $C^{\bullet}_{AS}(\widetilde{M})^{\Gamma}$.

We thus obtain the following exact sequence of cochain complexes

$$0 \longrightarrow C^{\bullet}_{AS,0}(\widetilde{M})^{\Gamma} \longrightarrow C^{\bullet}_{AS}(\widetilde{M})^{\Gamma} \longrightarrow \overline{C}^{\bullet}_{AS}(\widetilde{M})^{\Gamma} \longrightarrow 0 \tag{8.1}$$

where $\overline{C}^{\bullet}_{AS}(\widetilde{M})^{\Gamma}$ is the quotient complex. Let us denote in the following way the associated cohomology groups, forming the associated long exact sequence,

$$\ldots \longrightarrow H^{\bullet}_{AS,0,\Gamma}(\widetilde{M}) \longrightarrow H^{\bullet}_{AS,\Gamma}(\widetilde{M}) \longrightarrow \overline{H}^{\bullet}_{AS,\Gamma}(\widetilde{M}) \longrightarrow \ldots. \tag{8.2}$$

We first identify $H^{\bullet}_{AS,\Gamma}(\widetilde{M})$ with $H^{\bullet}(\Gamma)$. In order to do that fix a point z in $\widetilde{M}$ and consider the inclusion $i\colon \Gamma \hookrightarrow \widetilde{M}$ given by $i(\gamma) = z\gamma$. Observe that i is Γ-equivariant.

PROPOSITION 8.2. *The map* $i^*: C^\bullet_{AS}(\widetilde{M})^\Gamma \to C^\bullet_{AS}(\Gamma)^\Gamma$ *induces an isomorphism in cohomology and in fact is a chain homotopy equivalence.*

PROOF. Let us fix a bounded measurable fundamental domain $\mathcal{F}$ for the action of Γ on $\widetilde{M}$ such that $z \in \mathcal{F}$. Let $\mu\colon \widetilde{M} \to \Gamma$ be the map with $\mu(x) = \gamma$ for $x \in \mathcal{F}\gamma$, and notice that μ is Γ-equivariant. Observe that $\mu \circ i = \mathrm{id}_\Gamma$ and $r := i \circ \mu$ is the map which sends $x \in \mathcal{F}\gamma$ to $z\gamma$. To prove that i^* induces an isomorphism in cohomology we construct a cochain homotopy $K\colon C^k_{AS}(\widetilde{M})^\Gamma \to C^{k-1}_{AS}(\widetilde{M})^\Gamma$ between id and r^*, setting

$$K(\varphi)(x_0,\dots,x_{k-1}) := \sum_{j=0}^{k-1}(-1)^{j+1}\varphi(r(x_0),\dots,r(x_j),x_j,\dots,x_{k-1})$$

for $\varphi \in C^k_{AS}(\widetilde{M})^\Gamma$. Of course, since r is Γ-equivariant, so is K. We have

(8.3)

$$\begin{aligned}\delta K(\varphi)(x_0,\dots,x_k) &= \sum_{i=0}^{k}(-1)^i K(\varphi)(x_0,\dots,\hat{x}_i,\dots,x_k) = \\ &= \sum_{i=0}^{k}(-1)^i\left(\sum_{j=0}^{i-1}(-1)^{j+1}\varphi(r(x_0),\dots,r(x_j),x_j,\dots,\hat{x}_i,\dots,x_k)\right.\\ &\left.\quad+\sum_{j=i+1}^{k}(-1)^j\varphi(r(x_0),\dots,\widehat{r(x_i)},\dots,r(x_j),x_j,\dots,x_k)\right),\end{aligned}$$

(8.4)

$$\begin{aligned}K(\delta\varphi)(x_0,\dots,x_k) &= \sum_{j=0}^{k}(-1)^{j+1}\delta\varphi(r(x_0),\dots,r(x_j),x_j,\dots,x_k) = \\ &= \sum_{j=0}^{k}(-1)^{j+1}\left(\sum_{i=0}^{j}(-1)^i\varphi(r(x_0),\dots,\widehat{r(x_i)},\dots,r(x_j),x_j\dots,x_k)\right.\\ &\left.\quad+\sum_{i=j}^{k}(-1)^{i+1}\varphi(r(x_0),\dots,r(x_j),x_j,\dots,\hat{x}_i,\dots,x_k)\right).\end{aligned}$$

Comparing the terms of the two expressions we see that they cancel out perfectly, except for the term for $i, j = 0$ from the second sum which gives $-\varphi(x_0,\dots,x_k)$ and the term for $i, j = k$ from the second sum which gives $\varphi(r(x_0),\dots,r(x_k))$, which means that $\delta K + K\delta = r^* - \mathrm{id}$. The proposition follows. □

REMARK 8.3. We have the following variant of the cochain homotopy K in the proof of Proposition 8.2, which preserves continuous or smooth Alexander-Spanier cochains. Let $h\colon \widetilde{M} \to [0,1]$ be a compactly supported smooth cut-off function as in Section 6.2. Then define

$$\begin{aligned}&K\varphi(x_0,\dots,x_k)\\ &:= \sum_{\gamma_0,\dots,\gamma_k\in\Gamma^{k+1}} R^*_{\gamma_1}h(x_0)\cdots R^*_{\gamma_k}h(x_k)\sum_{j=0}^{k}(-1)^{j+1}\varphi(z\gamma_0,\dots,z\gamma_j,x_j,\dots,x_k)\end{aligned}$$

The construction in Proposition 8.2 is the same construction with h the characteristic function of the fundamental domain $\mathcal{F}$. The above argument again applies, using that $\sum_{\gamma\in\Gamma} R_\gamma^* h(x) = 1$ for each $x \in \widetilde{M}$.

REMARK 8.4. Notice that $C^\bullet_{AS}(\Gamma)^\Gamma$ is precisely the standard cobar complex used to define $H^\bullet(\Gamma; \mathbb{R})$.

Next we identify the remaining cohomology groups in (8.2) as certain singular cohomology groups. Let $B\Gamma$ be a classifying space for Γ, $E\Gamma \to B\Gamma$ a universal covering (a universal principal Γ-bundle). Let $\mu\colon M \to B\Gamma$ the map classifying the Γ-covering $\widetilde{M} \to M$, with Γ-equivariant lift $\tilde{\mu}\colon \widetilde{M} \to E\Gamma$. Using naturality of the constructions, we get a commutative diagram of cochain complexes

$$\begin{array}{ccc} C^\bullet_{AS}(\Gamma)^\Gamma & \xrightarrow{=} & C^\bullet_{AS}(\Gamma)^\Gamma \\ \uparrow{\scriptstyle\sim} & & \uparrow{\scriptstyle\sim} \\ C^\bullet_{AS}(E\Gamma)^\Gamma & \xrightarrow[\sim]{\tilde{\mu}^*} & C^\bullet_{AS}(\widetilde{M})^\Gamma \\ \downarrow{\scriptstyle f_E} & & \downarrow{\scriptstyle f_M} \\ \overline{C}^\bullet_{AS}(E\Gamma)^\Gamma & \xrightarrow{\overline{\tilde{\mu}}^*} & \overline{C}^\bullet_{AS}(\widetilde{M})^\Gamma \end{array}$$

The first vertical maps are induced by the inclusions $\Gamma \to \widetilde{M}$ and $\Gamma \to E\Gamma$ given by $g \mapsto zg$ and $g \mapsto \tilde{\mu}(z)g$ respectively. They are both chain homotopy equivalences by Proposition 8.2. Therefore also $\tilde{\mu}^*$ is a chain homotopy equivalence. For the quotients by the locally zero cochains, we note in addition that the pull-back defines canonical vertical isomorphisms

$$\begin{array}{ccc} \overline{C}^\bullet_{AS}(B\Gamma) & \xrightarrow{\overline{\mu}^*} & \overline{C}^\bullet_{AS}(M) \\ \downarrow{\scriptstyle\cong} & & \downarrow{\scriptstyle\cong} \\ \overline{C}^\bullet_{AS}(E\Gamma)^\Gamma & \xrightarrow{\overline{\tilde{\mu}}^*} & \overline{C}^\bullet_{AS}(\widetilde{M})^\Gamma \end{array} \tag{8.5}$$

Now, f_M is surjective with kernel $C^\bullet_{AS,0}(\widetilde{M})^\Gamma$. We get a sequence of cochain maps

$$\begin{aligned} C^\bullet_{AS,0}(\widetilde{M})^\Gamma \xrightarrow{\sim} C^\bullet(f_M) \xleftarrow[\tilde{\mu}^*]{\sim} C^\bullet(f_M \circ \tilde{\mu}^*) &= C^\bullet(\overline{\tilde{\mu}}^* \circ f_E) \\ &\xrightarrow{f_E} C^\bullet(\overline{\tilde{\mu}}^*) \cong C^\bullet(\overline{\mu}) =: C^\bullet_{AS}(M \to B\Gamma). \end{aligned} \tag{8.6}$$

The first map is induced by the inclusion and is the standard chain homotopy equivalence between the mapping cone of a surjection and the kernel. If the extension of chain complexes is $0 \to A^\bullet \xrightarrow{j} B^\bullet \xrightarrow{q} C^\bullet \to 0$, it sends $a \in A^n$ to $(j(a), 0) \in B^n \oplus C^{n-1} = C^n(q)$, which is the mapping cone complex. The second map is the standard map induced by g from the mapping cone of a composition $A^\bullet \xrightarrow{f} B^\bullet \xrightarrow{g} C^\bullet$ to the one of f: $C^n(g \circ f) = A^n \oplus C^{n-1} \ni (a, c) \mapsto (f(a), c) \in B^n \oplus C^{n-1} = C^n(g)$. It is standard and easy to see that this is a chain homotopy equivalence or homology isomorphism if and only if $f\colon A^\bullet \to B^\bullet$ is. Because of Proposition 8.2, the map induced by $\tilde{\mu}^*$ here is a chain homotopy equivalence (and implicitly, we have in mind to use a chain homotopy inverse of it). The next map, f_E, is obtained the same way. The isomorphism at the end implements the isomorphism (8.5). Finally, we use the definition of the relative cohomology of $M \to B\Gamma$

defined via Alexander-Spanier cochains. Because M is a manifold and $B\Gamma$ a CW-complex, this is canonically isomorphic to the relative singular cohomology.

PROPOSITION 8.5. *(Higson's Lemma) The sequence of cochain maps of* (8.6) *defines a canonical isomorphism*

$$H^\bullet_{AS,0,\Gamma}(\widetilde{M})^\Gamma \xrightarrow{\cong} H^\bullet(M \xrightarrow{u} B\Gamma).$$

PROOF. It only remains to prove that $f_E\colon C^\bullet_{AS}(E\Gamma)^\Gamma \to \overline{C}^\bullet_{AS}(E\Gamma)^\Gamma$ is a cohomology isomorphism. Note that the cohomology in both cases is the cohomology of Γ in a canonical way, and therefore there already is a canonical isomorphism between the two cohomology groups. We will explain this, and will show that f_E induces precisely this canonical isomorphism. This also makes the identification asserted in the proposition truly canonical.

We start with the free simplicial set generated by the vertices Γ, considered as a (discrete) set. Being free, this is contractible, and the associated chain complex $\dots\mathbb{Z}[\Gamma^3] \to \mathbb{Z}[\Gamma^2] \to \mathbb{Z}[\Gamma] \to \mathbb{Z}$ is acyclic. Note that it consists of free $\mathbb{Z}[\Gamma]$-modules, therefore is a free resolution of the trivial $\mathbb{Z}[\Gamma]$-module $\mathbb{Z}$. Note that our cochain equivalence with the bar complex $C^\bullet_{AS}(\Gamma)^\Gamma$ is obtained from a chain equivalence of the underlying free resolutions, and also $\tilde{\mu}^*$ is such a cochain equivalence.

To get the canonical identification of $\overline{H}^\bullet_{AS}(E\Gamma)^\Gamma$ with $H^\bullet(\Gamma)$, we use singular homology as intermediate step. Let $\Delta^\bullet(X)$ be the singular cochain complex of a space X. Let $\Delta^\bullet_0(X)$ be the subcomplex of cochains ϕ which are locally 0, i.e. by definition for which an open covering $\{U_i\}$ of X exists such that ϕ vanishes on all singular simplices $\sigma\colon \Delta^\bullet \to X$ whose image is contained in one of the U_i. Define $\overline{\Delta}^\bullet(X) := \Delta^\bullet(X)/\Delta^\bullet_0(X)$.

There is a canonical map $C^\bullet_{AS}(X) \to \Delta^\bullet(X)$ dual to the map between the singular chain complex $\Delta_\bullet(X)$ and the Alexander-Spanier chain complex $\mathbb{Z}[X^{\bullet+1}]$ which sends a singular simplex to the tuple of its endpoints. It induces $\overline{C}^\bullet_{AS}(X) \to \overline{\Delta}^\bullet(X)$.

It is a standard result, compare [**56**, Section 8.8], that for spaces with a nice local topology, in particular for CW-spaces like our $B\Gamma$ or M, these maps are chain homotopy equivalences, which we combine here with the projection induced isomorphisms with equivariant cochains on $E\Gamma$

$$\begin{array}{ccccc}
\overline{C}^\bullet_{AS}(B\Gamma) & \xrightarrow{\sim} & \overline{\Delta}^\bullet(B\Gamma) & \xleftarrow{\sim} & \Delta^\bullet(B\Gamma) \\
\downarrow{\cong} & & \downarrow{\cong} & & \downarrow{\cong} \\
\overline{C}^\bullet_{AS}(E\Gamma)^\Gamma & \xrightarrow{\sim} & \overline{\Delta}^\bullet(E\Gamma)^\Gamma & \xleftarrow{\sim} & \Delta^\bullet(E\Gamma)^\Gamma
\end{array} \tag{8.7}$$

Finally, as $E\Gamma$ is contractible, the singular chain complex of $E\Gamma$ provides another resolution of $\mathbb{Z}$ (by free $\mathbb{Z}\Gamma$-modules), and the canonical map to the Alexander-Spanier chain complex (which is a $\mathbb{Z}[\Gamma]$-module map) is automatically a chain equivalence. After taking $\mathrm{Hom}_{\mathbb{Z}[\Gamma]}(\cdot, \mathbb{R})$ this induces, once again, the standard cohomology isomorphism for different ways to compute group cohomology. Combining this information with (8.7) we obtain the diagram

$$\begin{array}{ccc}
C^\bullet_{AS}(E\Gamma)^\Gamma & \xrightarrow{\sim} & \Delta^\bullet(E\Gamma)^\Gamma \\
\downarrow{f_E} & & \downarrow{\sim} \\
\overline{C}^\bullet_{AS}(E\Gamma)^\Gamma & \xrightarrow{\sim} & \overline{\Delta}^\bullet(E\Gamma)^\Gamma.
\end{array}$$

It follows that our map f_E is a chain equivalence, and, because all the other maps induce the canonical isomorphisms, that it induces the canonical isomorphism between different ways to compute group cohomology. □

COROLLARY 8.6. *There exists a canonical and natural commutative diagram*

$$
\begin{array}{ccccccccc}
\dots \longrightarrow & H^\bullet_{AS,0,\Gamma}(\widetilde{M}) & \longrightarrow & H^\bullet_{AS,\Gamma}(\widetilde{M}) & \longrightarrow & \overline{H}^\bullet_{AS,\Gamma}(\widetilde{M}) & \longrightarrow \dots \\
& \downarrow & & \downarrow & & \downarrow & \\
\dots \longrightarrow & H^\bullet(M \xrightarrow{u} B\Gamma) & \longrightarrow & H^\bullet(B\Gamma) & \longrightarrow & H^\bullet(M) & \longrightarrow \dots
\end{array}
\tag{8.8}
$$

where the vertical lines are isomorphisms.

PROPOSITION 8.7. *In* (8.1), *the inclusion of the subcomplexes of smooth and antisymmetric cochains induce cohomology isomorphism. Here, a cochain* φ *is antisymmetric if* $\varphi(x_0, \dots, x_n) = \operatorname{sgn}(\sigma)\varphi(x_{\sigma(0)}, \dots, x_{\sigma(n)})$, *for all the permutations* $\sigma \in \mathfrak{S}_{n+1}$.

PROOF. By [**20**, Lemma 1.1 and Lemma 1.4], for calculating $\overline{H}^\bullet_{AS,\Gamma}(\widetilde{M})$ one can use the subcomplex of smooth antisymmetric Alexander-Spanier cochains φ. Analogously $H^\bullet(\Gamma)$ can be calculated by means of the subcomplex of antisymmetric group cochains c. This implies by the five lemma the same for $H^\bullet_{AS,0,\Gamma}(\widetilde{M})$. □

DEFINITION 8.8. Define the subcomplex $C^\bullet_{AS,0,pol}(\widetilde{M})^\Gamma$ of $C^\bullet_{AS,0}(\widetilde{M})^\Gamma$ consisting of smooth skew-symmetric cochains of polynomial growth, namely cochains φ such that

$$|\varphi(x_0, \dots, x_n)| \le K \prod_i (1 + d(x_i, x_{i+1}))^k$$

for some $K > 0$ and $k \in \mathbb{N}$.

PROPOSITION 8.9. *In our situation, the canonical inclusion induced map*

$$H^*(C^\bullet_{AS,0,pol}(\widetilde{M})^\Gamma) \to H^*(C^\bullet_{AS,0}(\widetilde{M})^\Gamma) \tag{8.9}$$

is an isomorphism if and only if $H^*_{pol}(\Gamma) \to H^*(\Gamma)$ *is an isomorphism. In particular,* (8.9) *is an isomorphism if* Γ *is a group with a polynomial combing, e.g. a hyperbolic group or more generally an automatic group, or if* Γ *has polynomial growth.*

PROOF. First, our proof of Proposition 8.2 works also for the subcomplexes of polynomial growth. This implies that $C^\bullet_{AS,pol}(\widetilde{M})^\Gamma \to C^\bullet_{AS}(\widetilde{M})^\Gamma$ is a cohomology isomorphism if and only if $C^\bullet_{AS,pol}(\Gamma)^\Gamma \to C^\bullet_{AS}(\Gamma)^\Gamma$ is one, i.e. if and only if $H^*_{pol}(\Gamma) \to H^*(\Gamma)$ is an isomorphism. Finally, we get a diagram with pullback isomorphisms

$$
\begin{array}{ccc}
\overline{C}^\bullet_{AS}(M) & \xrightarrow{\cong} & \overline{C}^\bullet_{AS,pol}(\widetilde{M})^\Gamma \\
\downarrow{=} & & \downarrow{i_{pol}} \\
\overline{C}^\bullet_{AS}(M) & \xrightarrow{\cong} & \overline{C}^\bullet_{AS}(\widetilde{M})^\Gamma,
\end{array}
$$

because in $\overline{C}$ only equivalence classes of cochains defined locally near the diagonal play a role, and locally the polynomial growth condition becomes a boundedness condition, but due to the compactness of M locally every smooth cochain automatically is bounded, so that indeed $\overline{C}^\bullet_{AS,pol}(\tilde{M})^\Gamma$ is isomorphic to $\overline{C}^\bullet_{AS}(M)$. The first statement then follows from the five lemma.

By [**60**, Corollary 5.3], if Γ admits a combing of polynomial growth, in particular if Γ is hyperbolic, then $H^*_{pol}(\Gamma) \to H^*(\Gamma)$ is an isomorphism. The same holds for groups of polynomial growth by [**59**]. This concludes the proof. □

8.2. Lifting $H^*(M \to B\Gamma)$ to relative cyclic cocycles

We continue to fix a smooth compact manifold M and a Γ-covering $\widetilde{M} \to M$.

LEMMA 8.10. *Given $\chi \in C^k_{\mathrm{AS},0}(\widetilde{M})^\Gamma$ smooth and antisymmetric and $A_0, \ldots, A_k \in \Psi^0_{\Gamma,c}(\widetilde{M}, \widetilde{E})$, then the (formal) expression*

(8.10)
$$\tau_\chi(A_0, \ldots, A_k) = \int_{\mathcal{F}} \mathrm{Tr}\left(\int_{\widetilde{M}^k} A_0(x_0, x_1) A_1(x_1, x_2) \ldots A_k(x_k, x_0) \chi(x_0, \ldots, x_k) \cdot dvol^k_g(x_1, \ldots, x_k) \right) dvol_g(x_0)$$

where $A_j(x, y)$ is the Schwartz kernel of A_j and $\mathcal{F}$ is a measurable fundamental domain for $\widetilde{M} \to M$

(1) *makes sense and gives a well defined value in $\mathbb{C}$ (independent of $\mathcal{F}$),*
(2) *vanishes if all the A_j are local operators, i.e. the support of the Schwartz kernel is supported on the diagonal (e.g. if the A_j are the operators of multiplication with a smooth function),*
(3) *is antisymmetric under cyclic permutations of the A_j,*
(4) *satisfies $\tau_{\partial\chi} = b\tau_\chi$ where ∂ is the Alexander-Spanier differential and b the cyclic differential.*

PROOF. To establish (1), by assumption, there exists $\delta > 0$ such that $\chi(x_0, \ldots, x_k) = 0$ if $d(x_i, x_j) < \delta$ for $i, j = 0, 1, \ldots, k$. Choose a smooth Γ-equivariant function $\alpha \colon \widetilde{M} \times \widetilde{M} \to [0, 1]$ which is 1 in a neighborhood of the diagonal and 0 outside the $\delta/2$-neighborhood of the diagonal. Set $\beta := 1 - \alpha$. To simplify the notation we assume $k = 1$, it will be clear how the argument generalizes.

We observe that $B_0(x, y) := A_0(x, y) \cdot \beta(x, y)$ is smooth and therefore a smoothing operator. The composition of B_0 with A_1 therefore is well defined and a smoothing operator with smooth kernel

$$(B_0 \circ A_1)(x_0, x_2) = \int_{\widetilde{M}} A_0(x_0, x_1) A_1(x_1, x_2) \beta(x_0, x_1)\, dx_1.$$

Note that the composition makes sense and the integral is over a compact subset of $\widetilde{M}$ due to the Γ-compactness of the supports of A_0 and A_1. If we multiply with $f(x_0)g(x_1)$ for smooth functions f, g, the kernel

$$(x_0, x_2) \mapsto \int_{\widetilde{M}} A_0(x_0, x_1) A_1(x_1, x_2) \beta(x_0, x_1) f(x_0) g(x_1)\, dx_1$$

is (by definition) the Schwartz kernel of the composition of m_f, B_0, m_g, A_1 which again is smoothing and indeed depends continuously on the function $(x, y) \mapsto f(x)g(y)$. Note that by the Γ-compact support condition, for fixed x_0, x_2 only the values of $f(x)g(y)$ on a compact subset of $\widetilde{M} \times \widetilde{M}$ enter. The span of functions of the form $(x, y) \mapsto f(x)g(y)$ is dense in all continuous functions of two variables and similarly for three variables. By continuity we therefore also define $\int_{\widetilde{M}} A_0(x_0, x_1) A_1(x_1, x_2) \beta(x_0, x_1) \chi(x_0, x_1, x_2)\, dx_1$. The operator $C_0 := A_0 - B_0$ has

Schwartz kernel $A_0(x,y)\alpha(x,y)$ with support in the $\delta/2$-neighborhood of the diagonal. Define $B_1(x,y) := A_1(x,y)\beta(x,y)$, again a smoothing kernel. As before, the function $(x_0,x_2) \mapsto \int_{\widetilde{M}} C_0(x_0,x_1)A_1(x_1,x_2)\beta(x_1,x_2)\chi(x_0,x_1,x_2)\,dx_1$ is well defined and smooth, obtained as Schwartz kernel of a composition which involves the smoothing operator B_1. Finally, observe that $\alpha(x_0,x_1)\alpha(x_1,x_2)\chi(x_0,x_1,x_2)$ is identically 0, as the product of the first two factors has support on the $\delta/2$-neighborhood of the diagonal where χ vanishes identically. Using

$$\begin{aligned}&\beta(x_0,x_1) - \beta(x_0,x_1)\beta(x_1,x_2) + \alpha(x_0,x_1)\alpha(x_1,x_2) + \beta(x_1,x_2)\\ =&\beta(x_0,x_1)\alpha(x_1,x_2) + \alpha(x_0,x_1)\alpha(x_1,x_2) + \beta(x_1,x_2) = \alpha(x_1,x_2) + \beta(x_1,x_2) = 1,\end{aligned}$$

we therefore define

$$\begin{aligned}&\int_{\widetilde{M}} A_0(x_0,x_1)A_1(x_1,x_2)\chi(x_0,x_1,x_2)\,dx_1\\ &\quad := \int_{\widetilde{M}} A_0(x_0,x_1)A_1(x_1,x_2)\beta(x_0,x_1)\chi(x_0,x_1,x_2)\,dx_1\\ &\quad\quad + \int_{\widetilde{M}} A_0(x_0,x_1)A_1(x_1,x_2)\beta(x_1,x_2)\chi(x_0,x_1,x_2)\,dx_1\\ &\quad\quad - \int_{\widetilde{M}} A_0(x_0,x_1)A_1(x_1,x_2)\beta(x_0,x_2)\beta(x_1,x_2)\chi(x_0,x_1,x_2)\,dx_1.\end{aligned}$$

The support condition on χ implies that this expression is independent of the chosen function α. By Γ-invariance, the final integral appearing in (8.10) is independent of $\mathcal{F}$ (indeed, the function descends to a smooth function of $x_0 \in M$).

Concerning (2) the formula clearly gives 0 if all the Schwartz kernels have support on the diagonal.

To prove (3), the reduction as in the proof of (1) shows that the expression which defines τ_χ reduces to the L^2-trace (the integral over $\mathcal{F}$ over the diagonal) of a composition of a smoothing Γ-equivariant operator with bounded Γ-invariant operators. We can then use the trace property of the L^2-trace [**1**] and get

$$\begin{aligned}&\tau_\chi(A_k, A_0,\dots,A_{k-1})\\ &= \int_{\mathcal{F}} \operatorname{Tr}\left(\int_{\widetilde{M}^k} A_k(x_0,x_1)A_0(x_1,x_2)\dots A_{k-1}(x_k,x_0)\chi(x_0,\dots,x_k)dx_1\cdots dx_k\right)dx_0\\ &= \int_{\mathcal{F}\times\widetilde{M}^k} \operatorname{Tr}\left(A_0(x_0,x_1)A_1(x_1,x_2)\dots A_k(x_k,x_0)\chi(x_k,x_0\dots,x_{k-1})\right)dx_0\cdots dx_k\end{aligned}$$

where we note that commuting the operators involves commuting the contribution of χ. Using the skew-adjointness $\chi(x_k,x_0,\dots,x_{k-1}) = (-1)^k\chi(x_0,\dots,x_k)$ gives the desired formula. The last property (4) is checked by a standard straightforward calculation, compare [**20**, Lemma 2.1]. □

COROLLARY 8.11. *Let $\mathfrak{m}_c\colon C^\infty(M) \to \Psi^0_{\Gamma,c}(\widetilde{M})$ be the inclusion as Γ-equivariant multiplication operators. Formula* (8.10) *defines a homomorphism*

$$H^\bullet(M \to B\Gamma) \xrightarrow{\cong} H^\bullet_{AS,0,\Gamma}(\tilde{M}) \to HC^\bullet(C^\infty(M) \xrightarrow{\mathfrak{m}_c} \Psi^0_{\Gamma,c}(\widetilde{M}))$$

given by $[\chi] \mapsto [(0,\tau_\chi)]$.

PROOF. By Lemma 8.10 (2) and (4), $\mathfrak{m}_c^*(\tau_\chi) = 0$, i.e. $(0,\tau_\chi)$ indeed is a relative cyclic cocycle, provided χ is a locally zero Alexander-Spanier cocycle, and

$(0, \tau_{\partial\chi}) = b(0, \tau_\chi)$. It follows that the assignment gives a well defined map between the homology groups. □

Similarly, we have an analogous homomorphism for non locally zero Alexander-Spanier cocycles. Indeed, by using exactly the same formula (8.10), we obtain the following.

COROLLARY 8.12. *There exists an homomorphism*

$$H^\bullet_{AS,\Gamma}(\widetilde{M}) \xrightarrow{\tau} HC^\bullet(0 \to \Psi^{-\infty}_{\Gamma,c}(\widetilde{M}))$$

given by $[\chi] \mapsto [(0, \tau_\chi)]$ *with* τ_χ *as in* (8.10).

Following Connes and Moscovici, let us recall, in an alternative fashion, the map from the Alexander-Spanier cohomology of M to the relative cyclic cohomology associated to π^*, $\overline{H}^\bullet_{AS}(M) \to HC^\bullet(C^\infty(M) \xrightarrow{\pi^*} \Sigma)$. Here Σ is the quotient of the 0-order pseudodifferential operator algebra by the ideal of smoothing operators (notice that we use the same notation for the quotient on $\widetilde{M}$ and the quotient on M because, as we have already seen, they are isomorphic). Let $\overline{\varphi}$ be an antisymmetric cocycle in $\overline{C}^k_{AS}(M)$ and let φ be any representative in $C^k_{AS}(M)$. If $A_0, A_1, \dots, A_k$ are such that at least one of them is a smoothing operator, then formula (8.10) gives cyclic cochain over $\Psi^{-\infty}(M)$, see [**20**, Lemma2.1]. Consider then the following two compositions

(8.11)
$$\partial^{HC}\colon HC^\bullet\left(\Psi^{-\infty}(M)\right) \longrightarrow HC^\bullet\left(\Psi^0(M) \xrightarrow{\sigma_{pr}} \Sigma\right) \xrightarrow{(\mathfrak{m},id)^*} HC^\bullet\left(C^\infty(M) \xrightarrow{\pi^*} \Sigma\right)$$

where $\mathfrak{m}\colon C(M) \to \Psi^0(M)$ is the obvious inclusion, and

(8.12)
$$\partial^{HC}_\Gamma\colon HC^\bullet\left(\Psi^{-\infty}_{\Gamma,c}(\widetilde{M})\right) \to HC^\bullet\left(\Psi^0_{\Gamma,c}(\widetilde{M}) \xrightarrow{\sigma_{pr}} \Sigma\right) \xrightarrow{(\mathfrak{m}_c,id)^*} HC^\bullet\left(C^\infty(M) \xrightarrow{\pi^*} \Sigma\right)$$

The first arrows in both compositions are given by excision, obtained as the inverse of the inclusion of smoothing operators into 0-order pseudodifferential operators.

DEFINITION 8.13. The Connes-Moscovici map

$$\overline{\tau}\colon \overline{H}^\bullet_{AS}(M) \to HC^\bullet\left(C^\infty(M) \xrightarrow{\pi^*} \Sigma\right)$$

is given by

$$[\overline{\varphi}] \mapsto \partial^{HC}(\tau_\varphi),$$

where τ_φ is formally defined as in (8.10).

For later use we point out the following compatibility result.

THEOREM 8.14. *The following diagram is commutative.*

(8.13)
$$\begin{array}{ccccccccc}
\longrightarrow & H^\bullet_{AS,\Gamma}(\widetilde{M}) & \longrightarrow & \overline{H}^\bullet_{AS}(M) & \xrightarrow{\partial^{AS}} & H^{\bullet+1}_{AS,0,\Gamma}(\widetilde{M}) & \longrightarrow \\
& \downarrow{\scriptstyle\tau} & & \downarrow{\scriptstyle\overline{\tau}} & & \downarrow{\scriptstyle\tau} & \\
\longrightarrow & HC^\bullet(\Psi^{-\infty}_{\Gamma,c}(\widetilde{M})) & \xrightarrow{\partial^{HC}_\Gamma} & HC^\bullet(\pi^*) & \xrightarrow{(1,\sigma_{pr})^*} & HC^\bullet(\mathfrak{m}_c) & \longrightarrow
\end{array}$$

PROOF. The commutativity of the first square is immediate: indeed, up to suspension of the target, the map $HC^{\bullet-1}(\mathfrak{m}_c) \to HC^{\bullet}(\Psi^{-\infty}_{\Gamma,c}(\widetilde{M}))$ is given by restriction along the inclusion of pairs $0 \to \Psi^{-\infty}_{\Gamma,c}(\widetilde{M})$ into $C^\infty(M) \xrightarrow{\mathfrak{m}_c} \Psi^0_{\Gamma,c}(\widetilde{M})$ and the commutativity follows directly by using explicit formulae.

The commutativity of the second square follows from the proof of [**20**, Theorem 5.2].

Finally, let us prove the commutativity of the third square. Consider a cocycle $\overline{\varphi}$ whose cohomology class belongs to $\overline{H}^*_{AS,\Gamma}(\widetilde{M})$. Let us recall the elementary construction of $\partial^{AS}[\overline{\varphi}]$ given by the snake lemma:

- by exactness of (8.1) there exists a cochain $\widetilde{\varphi} \in C^{\bullet}_{AS}(\widetilde{M})^\Gamma$ whose image in the quotient is $\overline{\varphi}$;
- consider then $\delta\widetilde{\varphi} \in C^{*+1}_{AS}(\widetilde{M})^\Gamma$: since $\overline{\varphi}$ is a cocycle it follows that the image of $\delta(\widetilde{\varphi})$ into the quotient complex is zero;
- it follow that $\delta\widetilde{\varphi}$ is the image of a locally zero cocycle $\psi \in C^{*+1}_{AS,0}(\widetilde{M})^\Gamma$, whose class represents $\partial^{AS}[\overline{\varphi}]$.

Let us now pick a relative cyclic cycle $[x, y] \in HC_*(\mathfrak{m}_c)$. In the following, with a small abuse of notation, we shall denote with α (instead of α_*) the map induced between cyclic complexes by an algebra homomorphism. Then we have that

$$\langle (0, \tau_{\partial^{AS}\overline{\varphi}}), (x, y)\rangle = \langle \tau_\psi, y\rangle = \langle \tau_{\delta\widetilde{\varphi}}, y\rangle = \langle \tau_{\widetilde{\varphi}}, \mathfrak{d} y\rangle$$

where $\mathfrak{d}$ denotes here the cyclic differential. Since (x, y) is a relative cycle, it follows that $\mathfrak{d} y = \mathfrak{m}_c(x)$ so

$$\langle \tau_{\widetilde{\varphi}}, \mathfrak{d} y\rangle = \langle \tau_{\widetilde{\varphi}}, \mathfrak{m}_c(x)\rangle.$$

Observe that $\widetilde{\varphi}$ is any lift of $\overline{\varphi}$, so we can mod out locally zero cochain in the left member of the pairing, transferring the pairing on M and not on $\widetilde{M}$, namely $\langle \tau_{\widetilde{\varphi}}, \mathfrak{m}_c(x)\rangle = \langle \tau_\varphi, \mathfrak{m}(x)\rangle$ where φ is a lift of $\overline{\varphi}$ in $\overline{C}^*_{AS}(M)$. Since $\sigma_{pr} \circ \mathfrak{m}(x) = \sigma_{pr} \circ \mathfrak{m}_c(x) = \sigma_{pr}(\mathfrak{d} y) = \mathfrak{d}(\sigma_{pr}(y))$, it follows that $(\mathfrak{m}(x), \sigma_{pr}(y))$ is a cycle for the pair $C(M) \xrightarrow{\mathfrak{m}} \Psi^0(M)$. Therefore we have that

$$\begin{aligned}\langle \tau_\varphi, \mathfrak{m}(x)\rangle &= \langle (\tau_\varphi, 0), (\mathfrak{m}(x), \sigma_{pr}(y))\rangle = \langle \tau_\varphi, \partial_{HC}(x, \sigma_{pr}(y))\rangle = \\ &= \langle \partial^{HC}(\tau_\varphi), (x, \sigma_{pr}(y))\rangle = \langle (1, \sigma_{pr})^* \circ \partial^{HC}(\tau_\varphi), (x, y)\rangle,\end{aligned} \tag{8.14}$$

where ∂_{HC} is the map in cyclic homology dual to (8.11) and the second equality is obtained by excision. Summarizing, we have proved that

$$(1, \sigma_{pr})^* \overline{\tau}_{[\overline{\varphi}]} = \tau_{\partial^{AS}[\overline{\varphi}]},$$

namely the commutativity of the third square. □

8.3. Rapid decay

We next need a different kind of smooth subalgebra of pseudodifferential operators. In this section we shall assume that Γ has property (RD). We recall the definition below. Consider the Lie groupoid $G := \widetilde{M} \times_\Gamma \widetilde{M} \rightrightarrows M$. Let us recall some basic definitions.

DEFINITION 8.15. A length function on G is a proper and continuous function $l\colon G \to \mathbb{R}_+$ such that $l(\gamma\gamma') \leq l(\gamma) + l(\gamma')$, $l(\gamma) = l(\gamma^{-1})$, and $l(x) = 0$ for all $x \in M$ and $\gamma, \gamma' \in G$ composable.

EXAMPLE 8.16. Let d be any proper Γ-invariant distance function on $\widetilde{M}$. Let $[\widetilde{x}, \widetilde{y}]$ be an element of G. Then $l([\widetilde{x}, \widetilde{y}]) := d(\widetilde{x}, \widetilde{y})$ is a proper length function on G.

Let us fix a non-zero length function l on G. Define for $m \in \mathbb{N}$ the following norms on $C_c^\infty(\widetilde{M} \times_\Gamma \widetilde{M})$:

$$\|f\|_{m,l}^2 = \max\left\{ \sup_{x\in M} \int_{s^{-1}(x)} |f(\tilde{x}, \tilde{y})|^2 |1 + l(\tilde{x}, \tilde{y})|^{2m} , \right. \tag{8.15}$$
$$\left. \sup_{y\in M} \int_{r^{-1}(y)} |f(\tilde{x}, \tilde{y})|^2 |1 + l(\tilde{x}, \tilde{y})|^{2m} \right\}.$$

DEFINITION 8.17. Set $S_l^2(\widetilde{M} \times_\Gamma \widetilde{M}) := \mathrm{invlim}_{m\in\mathbb{N}} L_{m,l}$, where $L_{m,l}$ is the completion of $C_c^\infty(\widetilde{M} \times_\Gamma \widetilde{M})$ with respect to the norm $\|\cdot\|_{m,l}$. We say that G has property (RD) with respect to l if there exists $m \in \mathbb{N}$ and $C > 0$ such that

$$\|f\|_{red} \leq C \|f\|_{l,m} \quad \forall f \in C_c^\infty(\widetilde{M} \times_\Gamma \widetilde{M}),$$

where $\|\cdot\|_{red}$ denotes the norm of the reduced groupoid C*-algebra.

Recall that, by definition, the finitely generated group Γ itself has property (RD) if it has property (RD) as a groupoid over $\{*\}$, i.e. if $G = \Gamma \times_\Gamma \Gamma \rightrightarrows \{*\}$ has. In other words, there must exist $C > 0$ and $m \in \mathbb{N}$ such that $\|f\|_{red}^2 \leq C \sum_{g\in\Gamma} |f(g)|^2 (1 + l(g))^{2m}$ for all $f \in \mathbb{C}[\Gamma]$, where $l(\cdot)$ is the word length function associated to a finite set of generators.

PROPOSITION 8.18. *If the finitely generated group Γ has property (RD) then $\widetilde{M} \times_\Gamma \widetilde{M}$ is also (RD) with respect to the length function $l\colon [\widetilde{x}, \widetilde{y}] \mapsto d(\widetilde{x}, \widetilde{y})$, where the metric d on $\widetilde{M}$ is induced from a Riemannian metric on M via pullback.*

PROOF. Let $\mathcal{F}$ be a bounded measurable fundamental domain for the action of Γ on $\widetilde{M}$. Fix $u \in L^2(\widetilde{M})$ and $f \in C_c^\infty(\widetilde{M} \times_\Gamma \widetilde{M})$. We also use the letter f for its Γ-equivariant pullback to $\widetilde{M} \times \widetilde{M}$. Then

$$
\begin{aligned}
|f * u|^2_{L^2(\widetilde{M})} &= \sum_{g\in\Gamma} \int_{\mathcal{F}g} \left| \sum_{h\in\Gamma} \int_{\mathcal{F}h} f(\tilde{x},\tilde{y})u(\tilde{y})\, d\tilde{y} \right|^2 d\tilde{x} \\
&\le \sum_{g\in\Gamma} \int_{\mathcal{F}} \left| \sum_{h\in\Gamma} \int_{\mathcal{F}} |f(\tilde{x}g,\tilde{y}h)|\, |u(\tilde{y}h)|\, d\tilde{y} \right|^2 d\tilde{x} \\
&\le \int_{\mathcal{F}} \sum_{g\in\Gamma} \Big| \sum_{h\in\Gamma} \underbrace{\left(\int_{\mathcal{F}} \Big| \underbrace{f(\tilde{x}g,\tilde{y}h)}_{=f(\tilde{x},\tilde{y}hg^{-1})} \Big|^2 d\tilde{y} \right)^{1/2}}_{=:vf(\tilde{x},gh^{-1})} \underbrace{\left(\int_{\mathcal{F}} |u(\tilde{y}h)|^2\, d\tilde{y} \right)^{1/2}}_{=:vu(h)} \Big|^2 d\tilde{x} \\
&= \int_{\mathcal{F}} \sum_{g\in\Gamma} |vf(\tilde{x},\cdot) * vu(\cdot)(g)|^2 \\
&\le \int_{\mathcal{F}} \|vf(\tilde{x},\cdot)\|^2_{red} \cdot \sum_{h\in\Gamma} |vu(h)|^2\, d\tilde{x} \\
&\le C \int_{\mathcal{F}} \sum_{g\in\Gamma} |vf(\tilde{x},g)|^2 (1+l(g))^{2m} \cdot \left(\sum_{h\in\Gamma} \int_{\mathcal{F}} |u(\tilde{y}h)|^2\, d\tilde{y} \right) \\
&\le CC' \underbrace{\int_{\mathcal{F}} \sum_{g\in\Gamma} \int_{\mathcal{F}} |f(\tilde{x},\tilde{y}g)|^2 (1+d(\tilde{x},\tilde{y}g))^{2m}\, d\tilde{y}\, d\tilde{x}}_{=\int_{x\in M}\int_{s^{-1}(x)} |f(\tilde{x},\tilde{y})|^2 (1+d(\tilde{x},\tilde{y}))^{2m}} \cdot |u|^2_{L^2(\widetilde{M})} \\
&\le CC' \operatorname{vol}(M) \underbrace{\sup_{x\in M} \int_{s^{-1}(x)} |f(\tilde{x},\tilde{y})|^2 (1+d(\tilde{x},\tilde{y}))^{2m}}_{\|f\|^2_{m,l}} \cdot |u|^2_{L^2(\widetilde{M})}
\end{aligned}
$$

The first inequality uses the elementary fact that the L^2-norm is unconditional (i.e. the norm of a convolution does never decrease if the functions are replaced by their absolute value functions). The second inequality is the Cauchy-Schwartz inequality for the integral over $\mathcal{F}$.

Then we use that

$$
\begin{aligned}
h \mapsto vu(h) &= \int_{\mathcal{F}} \left(|u(\tilde{y}h)|^2\, d\tilde{y} \right)^{1/2} \in l^2(\Gamma); \\
g \mapsto vf(\tilde{x},g) &= \left(\int_{\mathcal{F}} |f(\tilde{x},\tilde{y}g^{-1})|^2\, d\tilde{y} \right)^{1/2} \in \mathbb{C}[\Gamma].
\end{aligned}
$$

The forth inequality is property (RD) for Γ. Finally, we use the Milnor-Svarc result that the map $\Gamma \to \widetilde{M}$ given by $g \mapsto \tilde{x}g$ is a quasi-isometry, with constants uniform for x in the compact subset $\overline{\mathcal{F}}$, to obtain the fifth inequality. As C, C' do not depend on f and u, we have proved that $\widetilde{M} \times_\Gamma \widetilde{M}$ has property (RD). □

PROPOSITION 8.19. *The *-algebra $S^2_l(\widetilde{M} \times_\Gamma \widetilde{M})$ is dense and holomorphically closed in $C^*_{red}(\widetilde{M} \times_\Gamma \widetilde{M})$.*

PROOF. The proof of the corresponding result [**35**, Theorem 4.2] for r-discrete groupoids also works in this case, replacing sums over the r-fibers by integrals. This follows an idea which already appeared in [**20**], and more concretely in [**80**, Section 4]. □

DEFINITION 8.20. Set

$$S_l^{2,\infty} := \{f \in S_l^2 \,:\, f \text{ is smooth and } \|f\|_{m,l,\alpha,\beta} < \infty \quad \forall m, \alpha, \beta\},$$

where $\|f\|^2_{m,l,\alpha,\beta}$ is defined as

$$\max\left\{ \sup_{x \in M} \int_{s^{-1}(x)} |\nabla_x^\alpha \nabla_y^\beta f(\tilde{x},\tilde{y})|^2 |1 + l(\tilde{x},\tilde{y})|^{2m}, \right.$$

$$\left. \sup_{y \in M} \int_{r^{-1}(y)} |\nabla_x^\alpha \nabla_y^\beta f(\tilde{x},\tilde{y})|^2 |1 + l(\tilde{x},\tilde{y})|^{2m} \right\}.$$

PROPOSITION 8.21. *The *-algebra* $S_l^{2,\infty}$ *is dense and holomorphically closed in* $C^*_{red}(\widetilde{M} \times_\Gamma \widetilde{M})$.

PROOF. It is immediate that $S_l^{2,\infty}$ is a dense semi-ideal in S_l^2. Then we use [**43**, Prop 3.3] to conclude. □

PROPOSITION 8.22. *Denote by* $\Psi^0_{\Gamma,rd}(\widetilde{M})$ *the sum* $\Psi^0_{\Gamma,c}(\widetilde{M}) + S_l^{2,\infty}(\widetilde{M} \times_\Gamma \widetilde{M})$. *It is a *-algebra and it is dense and holomorphically closed in* $\Psi^0_\Gamma(\widetilde{M})$.

PROOF. The proof of this proposition follows verbatim the proof of [**43**, Corollary 7.9]. It is enough to replace $\mathcal{G}$, $\Psi^0(\mathcal{G})$, $\mathcal{S}(\mathcal{G}, \phi)$ and $\Psi^0_s(\mathcal{G})$ there with $\widetilde{M} \times_\Gamma \widetilde{M}$, $\Psi^0_{\Gamma,c}(\widetilde{M})$, $S_l^{2,\infty}(\widetilde{M} \times_\Gamma \widetilde{M})$ and $\Psi^0_{\Gamma,rd}(\widetilde{M})$, respectively. □

8.4. Extending cyclic cocycles associated to classes in $H^*_{pol}(M \xrightarrow{u} B\Gamma)$

Let $\chi \colon \widetilde{M^{k+1}} \to \mathbb{R}$ be a smooth anti-symmetric k-dimensional delocalized Alexander-Spanier cocycle in $C^\bullet_{AS,0}(\widetilde{M})^\Gamma$ of polynomial growth. Take the corresponding cyclic cocycle τ_χ as in Lemma 8.10 and Corollary 8.11, defined on the algebra $\Psi^0_{\Gamma,c}(\widetilde{M})$. In this section we are going to prove that χ extends continuously to the algebra $\Psi^0_{\Gamma,rd}(\widetilde{M})$. Recall that this algebra is defined as the sum of $\Psi^0_{\Gamma,c}(\widetilde{M})$ and $S_l^{2,\infty}(\widetilde{M} \times_\Gamma \widetilde{M})$ of Definition 8.20. Hence it is sufficient to prove the following lemma.

LEMMA 8.23. *The cyclic cocycle* τ_χ *extends continuously to the algebra*

$$S_l^{2,\infty}(\widetilde{M} \times_\Gamma \widetilde{M}).$$

PROOF. Indeed it is sufficient to prove that it extends to $S_l^2(\widetilde{M} \times_\Gamma \widetilde{M})$. To this end we have to show that τ_χ is continuous with respect to some seminorm of the type (8.15). We shall apply the method used in [**20**, Proposition 6.5]. By the polynomial growth of χ we know that there exists a constant $C > 0$ and an integer $n \geq 0$ such that

$$|\chi(x_0, x_1, \ldots, x_k)| \leq C(1 + d(x_0, x_1))^{2n} \ldots (1 + d(x_{k-1}, x_k))^{2n} \tag{8.16}$$

for all $x_0, x_1, \dots, x_k \in \widetilde{M}$.

Let $A_0, \dots, A_k$ be elements in $C_c^\infty(\widetilde{M} \times_\Gamma \widetilde{M})$ and put $B_i(x,y) := |A_i(x,y)|(1 + d(x,y))^{2n}$ for $i = 0, \dots, k-1$ and $B_k(x,y) := |A_k(x,y)|$. Then

(8.17)
$$
\begin{aligned}
&|\tau_\chi(A_0, \dots, A_k)| \\
&\quad = \left| \int_{\mathcal{F}} Tr \left(\int_{\widetilde{M}^k} A_0(x_0,x_1) \cdots A_x(x_k,x_0) \chi(x_0, \dots, x_k)\, dx_1 \dots dx_k \right) dx_0 \right| \\
&\quad \le C \int_{\mathcal{F}} \left(\int_{\widetilde{M}^k} B_0(x_0,x_1) B_1(x_1,x_2) \dots B_k(x_k,x_0) dx_1 \dots dx_k \right) dx_0 \\
&\quad = C \int_{\mathcal{F}} (B_0 * B_1 * \cdots * B_k)(x_0,x_0) dx_0 \\
&\quad \le C' \sup_{x \in \mathcal{F}} |B_0 * B_1 * \cdots * B_k(x,x)| \\
&\quad \le C'' \|B_0 * B_1 * \cdots * B_k\|_{red} \\
&\quad \le C'' \|B_0\|_{red} \|B_1\|_{red} \cdots \|B_k\|_{red}
\end{aligned}
$$

where in the last but one step we use the fact that the map $C^*_{red}(G) \to C_0(G^{(0)})$, given by the restriction to the units, is a continuous conditional expectation of C*-algebras.
Then, since the groupoid has property (RD), we have that

$$
\begin{aligned}
|\tau_\chi(A_0, \dots, A_k)| &\le C''' \|B_0\|_{l,m} \|B_1\|_{l,m} \cdots \|A_k\|_{red} \\
&\le C''' \|A_0\|_{l,m+n} \|A_1\|_{l,m+n} \cdots \|A_k\|_m
\end{aligned}
\tag{8.18}
$$

which gives the desired result. □

Propositions 8.5 and 8.9, Lemma 8.23, and Corollary 8.11 now give the following result.

THEOREM 8.24. *Let Γ be a discrete group with property (RD). Let $\widetilde{M} \to M$ be a Γ-covering of a compact manifold M. Then there is a well-defined homomorphism*

$$\Xi \colon H^*_{pol}(M \to B\Gamma) \to HC^*(C^\infty(M) \xrightarrow{\mathfrak{m}} \Psi^0_{\Gamma,rd}(\widetilde{M})).$$

Using Proposition 8.22 we then get a well defined pairing

$$H^*_{pol}(M \to B\Gamma) \times K_{*+1}\left(C(M) \to \Psi^0_\Gamma(\widetilde{M})\right) \to \mathbb{C},$$

explicitly given by associating to (α, x) the number $\langle \Xi(\alpha), x \rangle$, where the pairing between cyclic cohomology and K-theory has been used.

If, in addition, $H^*_{pol}(\Gamma) \to H^*(\Gamma)$ is an isomorphism, we can replace $H^*_{pol}(M \to B\Gamma)$ by the usual relative cohomology $H^*(M \to B\Gamma)$. Examples of groups which satisfy both conditions are hyperbolic groups or groups of polynomial growth [**59, 60**].

COROLLARY 8.25. *There exists a commutative diagram of the following shape:*

$$
\begin{array}{ccccccccc}
\cdots & \longrightarrow & H^*_{pol}(B\Gamma) & \longrightarrow & H^*(M) & \xrightarrow{\delta} & H^{*+1}_{pol}(M \to B\Gamma) & \longrightarrow \\
 & & \downarrow & & \downarrow & & \downarrow & \\
\cdots & \longrightarrow & HC^*(\Psi^{-\infty}_{\Gamma,rd}(\widetilde{M})) & \xrightarrow{\partial^{HC}_\Gamma} & HC^*(\pi^*) & \xrightarrow{(1,\sigma_{pr})^*} & HC^*(\mathfrak{m}_{rd}) & \longrightarrow
\end{array}
\tag{8.19}
$$

Here, $\mathfrak{m}_{rd}$ *is the inclusion as multiplication operators* $C^\infty(M) \to \Psi^0_{\Gamma,rd}(\widetilde{M})$ *and* π^* *is the composition of* $\mathfrak{m}_{rd}$ *and the quotient map to* Σ.

DEFINITION 8.26. Let Γ be a discrete group with property (RD). Let $\widetilde{D}$ be a generalized Dirac operator which is Γ-equivariant on the Galois Γ-covering $\widetilde{M}$ of a compact smooth manifold M. Suppose that $\widetilde{D}$ is L^2-invertible. Then define its higher rho number associated to $[\alpha] \in H^{[*-1]}_{pol}(M \to B\Gamma)$ as

$$\varrho_\alpha(\widetilde{D}) := \left\langle \Xi(\alpha), \varrho(\widetilde{D}) \right\rangle \in \mathbb{C},$$

where $\varrho(\widetilde{D})$ is defined as in Definition 3.4.

DEFINITION 8.27. Let g be a metric with positive scalar curvature on a closed spin manifold M of dimension n with fundamental group Γ. We define the higher rho number associated to g and to $\alpha \in H^{[*-1]}_{pol}(M \to B\Gamma)$ as

$$\varrho_\alpha(g) := \langle \Xi(\alpha), \varrho(g) \rangle \in \mathbb{C},$$

where $\varrho(g) \in K_n\left(C(M) \to \Psi^0_\Gamma(\widetilde{M})\right)$ is the rho class of the spin Dirac operator for the metric g.

CHAPTER 9

Delocalized higher Atiyah-Patodi-Singer indices

Let us consider as in Section 6.5 a cocompact even-dimensional Γ-covering with boundary $\widetilde{W} \to W$ with product metric near the boundary, and a Γ-equivariant Dirac operator $\widetilde{D}_W$ on $\widetilde{W}$ such that $\partial\widetilde{W} = \widetilde{M}$ and such that the boundary operator of $\widetilde{D}_W$ is equal to $\widetilde{D}_M$. Moreover, let us assume that Γ is Gromov hyperbolic or of polynomial growth. In this section we discuss how the higher rho numbers we have defined in the previous sections enter into index formulae for delocalized higher APS indices.

9.1. Higher APS indices associated to elements in $HC^*(\mathbb{C}\Gamma, \langle x\rangle)$

Let $\mathrm{Ind}_{\Gamma,b}(\widetilde{D}_W)$ be the class in $K_*(\mathcal{A}\Gamma)$ from Theorem 6.29. Let us fix a delocalized cyclic cocycle τ of polynomial growth such that $[\tau] \in HP^{even}_{del}(\mathbb{C}\Gamma)$. Then define

$$\mathrm{Ind}^{\tau}_{\Gamma,b}(\widetilde{D}_W) := \langle \mathrm{Ch}_\Gamma(\mathrm{Ind}_{\Gamma,b}(\widetilde{D}_W)), \tau\rangle \in \mathbb{C} \tag{9.1}$$

where the pairing is defined in the following way.

- $\mathrm{Ch}_\Gamma(\mathrm{Ind}_{\Gamma,b}(\widetilde{D}_W))$ is an element in $H_*(\mathcal{A}\Gamma)$.
- Consider its image in $\overline{HC}_*(\mathcal{A}\Gamma)$ using Corollary 7.19, see (7.9). With a small abuse of notation we keep the notation $\mathrm{Ch}_\Gamma(\mathrm{Ind}_{\Gamma,b}(\widetilde{D}_W))$ for this element of $\overline{HC}_*(\mathcal{A}\Gamma)$.
- Now, by Proposition 7.18, $\overline{HC}_*(\mathcal{A}\Gamma)$ splits as $\overline{HC}^e_*(\mathcal{A}\Gamma) \oplus HC^{del}_*(\mathcal{A}\Gamma)$ and then, following (6.38), we have that
$$\mathrm{Ch}_\Gamma(\mathrm{Ind}_{\Gamma,b}(\widetilde{D}_W)) = \pi^e\left(\mathrm{Ch}_\Gamma(\mathrm{Ind}_{\Gamma,b}(\widetilde{D}_W))\right) \oplus \pi^{del}\left(\mathrm{Ch}_\Gamma(\mathrm{Ind}_{\Gamma,b}(\widetilde{D}_W))\right),$$
where the first summand is an element of the reduced cyclic homology group localized at the identity and the second summand lies in unreduced $HC^{del}_*(\mathcal{A}\Gamma)$.

We *define* $\langle \mathrm{Ch}_\Gamma(\mathrm{Ind}_{\Gamma,b}(\widetilde{D}_W)), \tau\rangle =: \mathrm{Ind}^{\tau}_{\Gamma,b}(\widetilde{D}_W)$ as the pairing of

$$\pi^{del}\left(\mathrm{Ch}_\Gamma(\mathrm{Ind}_{\Gamma,b}(\widetilde{D}_W))\right) \in HC^{del}_*(\mathcal{A}\Gamma)$$

with $[\tau] \in HC^*_{del}(\mathbb{C}\Gamma)$.

Thanks to (6.38) and (6.39), we obtain the following equality

$$\mathrm{Ind}^{\tau}_{\Gamma,b}(\widetilde{D}_W) = -\frac{1}{2}\langle \varrho_{Lott}(\widetilde{D}_M), \tau\rangle \in \mathbb{C},$$

where $\varrho_{Lott}(\widetilde{D}_M)$ is the higher rho class of Lott, which we pair with τ.

In the even dimensional case we also have, by Proposition 6.32,

$$\pi^{del}\left(\mathrm{Ch}_\Gamma(\mathrm{Ind}_{\Gamma,b}(\widetilde{D}_W))\right) = \mathrm{Ch}^{del}_\Gamma\left(\varrho(\widetilde{D}_M)\right) \in H^{del}_{even}(\mathcal{A}\Gamma),$$

so that we obtain the following formula for the higher delocalized APS indices

$$\mathrm{Ind}^{\tau}_{\Gamma,b}(\widetilde{D}_W) = \varrho_\tau(\widetilde{D}_M) \in \mathbb{C}.$$

Summarizing, in the even dimensional case, for the higher delocalized APS indices we have the following double equality of complex numbers:

$$-\frac{1}{2}\langle \varrho_{Lott}(\widetilde{D}_M), \tau\rangle = \mathrm{Ind}^{\tau}_{\Gamma,b}(\widetilde{D}_W) = \varrho_\tau(\widetilde{D}_M) \tag{9.2}$$

If we apply all this to the delocalized trace associated to $\langle x\rangle$, that is

$$\tau_{\langle x\rangle}(\sum \alpha_\gamma \gamma) := \sum_{\gamma\in\langle x\rangle} \alpha_\gamma \gamma\,,$$

we obtain from Example 7.27 the following formula:

$$\mathrm{Ind}^{\tau_{\langle x\rangle}}_{\Gamma,b}(\widetilde{D}_W) = -\frac{1}{2}\eta_{\langle x\rangle}(\widetilde{D}_M) \tag{9.3}$$

with $\eta_{\langle x\rangle}(\widetilde{D}_M)$ denoting the delocalized eta invariant of Lott.

REMARK 9.1. Notice that equality (9.3) has been known for quite some time and it also holds for a general group Γ once we assume that $\langle x\rangle$ has polynomial growth: it is a direct consequence of the specialization to 0-degree of the higher APS index theorem in non-commutative de Rham homology (Theorem 6.29). This is discussed in detail in [**63**] for arbitrary Γ and $\langle x\rangle$ of polynomial growth, but the argument given there also applies to hyperbolic groups, once we use *directly* the results of Puschnigg [**68**].

9.2. Higher APS indices associated to elements in $H^*(M \to B\Gamma)$

Let χ be a cocycle defining a class in $H^{\Gamma}_{AS,0}(\widetilde{W})$. Consider the image of $[\chi]$ through the map

$$H^{even}_{AS,0,\Gamma}(\widetilde{W}) \xrightarrow{\alpha} H^{even}_{AS,\Gamma}(\widetilde{W}).$$

By Theorem 8.14, we have that

$$\mathrm{Ind}^{\alpha(\chi)}_{\Gamma,b}(\widetilde{D}_W) := \langle \tau_{\alpha(\chi)}, \mathrm{Ind}_{\Gamma,b}(\widetilde{D}_W)\rangle = \langle \tau_\chi, j_*\,\mathrm{Ind}_{\Gamma,b}(\widetilde{D}_W)\rangle,$$

where $j_*\colon K_*(\mathcal{S}^{2,\infty}_l(\overset{\circ}{\widetilde{W}} \times_\Gamma \overset{\circ}{\widetilde{W}})) \to K_*(\Psi^0_{\Gamma,rd}(\overset{\circ}{\widetilde{W}}))$. Since χ is a delocalized cocycle, reasoning as in (6.48), we have that

$$\langle \tau_\chi, j_*\,\mathrm{Ind}_{\Gamma,b}(\widetilde{D}_W)\rangle = \langle \tau_\chi, [\pi_\geq]\otimes e\rangle \tag{9.4}$$

where with a small abuse of notation we identify $[\pi_\geq]\otimes e$ with its image in the group $K_*(\Psi^0_{\Gamma,rd}(\overset{\circ}{\widetilde{W}}))$. Now, since $\pi_\geq \otimes e$ is supported in a open collar neighbourhood $\partial\widetilde{W} \times (0,1)$ of the boundary, we can easily see, from the explicit expression (8.10), that the right hand side of (9.4) is equal to $\langle \tau_{\iota^*\chi}, [\pi_\geq]\otimes e\rangle$ with ι the natural inclusion of $\partial\widetilde{W} \times (0,1)$ into $\widetilde{W}$. Summarizing,

$$\langle \tau_{\alpha(\chi)}, \mathrm{Ind}_{\Gamma,b}(\widetilde{D}_W)\rangle = \langle \tau_{\iota^*\chi}, [\pi_\geq]\otimes e\rangle\,.$$

By the Künneth theorem, we can assume that $\iota^*\chi$ is the product of the restriction $\iota^*_\partial\chi$ of the cocycle to $\partial\widetilde{W}$ and the generator of the cohomology of the interval. Then, we finally obtain the following delocalized APS-type equality

$$\mathrm{Ind}^{\alpha(\chi)}_{\Gamma,b}(\widetilde{D}_W) = \varrho_{\iota^*_\partial\chi}(\widetilde{D}_M) \text{ in } \mathbb{C}. \tag{9.5}$$

9.3. Filling of positive scalar curvature metrics

From the above formulae we immediately obtain the following result.

Let M be a closed compact n-dimensional spin manifold with a classifying map $f : M \to B\Gamma$ defining a Galois Γ-covering $\widetilde{M}$. Let g_M be a metric of positive scalar curvature on M. Assume now that (M, f) is null-bordant in $\Omega_n^{\mathrm{spin}}(B\Gamma)$; thus there exists a compact spin manifold W and $F\colon W \to B\Gamma$ such that $\partial W = M$ and $F|_\partial = f$.

PROPOSITION 9.2. *If g_M extends to a positive scalar curvature metric g_W, product-type near the boundary, then all higher rho numbers on $\Gamma \to \widetilde{M} \to M$ vanish.*

Of course, we make the necessary assumptions on Γ so that these higher rho numbers are indeed well-defined.

CHAPTER 10

The space of metrics of positive scalar curvature and higher rho-invariants

The techniques developed in this paper have interesting geometric applications. More specifically, we can use them to study important aspects of the topology of positive scalar curvature. The basic question we want to answer is: what can one say about the space of Riemannian metrics of positive scalar curvature on M? Let us start with a definition to fix some notation.

DEFINITION 10.1. Let M be a fixed smooth compact manifold without boundary of dimension $n \geq 5$.

- Let $\mathrm{Riem}^+(M)$ *be the space of Riemannian metrics on* M *with positive scalar curvature* (with its usual C^∞-topology).
- Let us denote its set of path components by $\pi_0(\mathrm{Riem}^+(M))$.
- The set $\mathcal{P}^+(M)$ *denotes the set of concordance classes of metrics of positive scalar curvature.* Two metrics $g_0, g_1 \in \mathrm{Riem}^+(M)$ are defined to be *concordant* if there is a metric G on $M \times [0,1]$ which has positive scalar curvature, is of product structure near the boundary, and restricts to g_0 on $M \times \{0\}$ and to g_1 on $M \times \{1\}$.

It is a standard fact that two metrics in the same path component of $\mathrm{Riem}^+(M)$ are also concordant, i.e. $\mathcal{P}^+(M)$ is a quotient of $\pi_0(\mathrm{Riem}^+(M))$.

A central tool for the study of positive scalar curvature metrics is Stolz' positive scalar curvature exact sequence (where $\Gamma = \pi_1(M)$)
(10.1)

$$\cdots \longrightarrow \Omega^{\mathrm{spin}}_{n+1}(M) \longrightarrow \mathrm{R}^{\mathrm{spin}}_{n+1}(\Gamma) \xrightarrow{\partial} \mathrm{Pos}^{\mathrm{spin}}_n(M) \longrightarrow \Omega^{\mathrm{spin}}_n(M) \longrightarrow \cdots$$

Again, to fix notation let us recall the definition of the terms in the Stolz positive scalar curvature exact sequence.

The three groups appearing in the Stolz sequence are bordism groups. In each case, cycles of the k-th group are k-dimensional manifolds X with extra structure and with a map $f\colon X \to M$. For Ω^{spin}_k, the manifolds X are closed and the extra structure is a spin structure. For $\mathrm{Pos}^{\mathrm{spin}}_k$, the cycles are closed manifolds and the extra structure is a spin structure together with a metric g of positive scalar curvature on X. Finally, for $\mathrm{R}^{\mathrm{spin}}_k$, the cycles are compact manifolds with boundary, the extra structure is a spin structure and a metric of positive scalar curvature on the boundary. For all of them the group structure is given by disjoint union.

REMARK 10.2. The type of cycles we have given a priori defines a group $\mathrm{R}^{\mathrm{spin}}_k(M)$. Post-composing with a map $u\colon M \to B\Gamma$ classifying a universal covering gives a canonical isomorphism $\mathrm{R}^{\mathrm{spin}}_k(M) \to \mathrm{R}^{\mathrm{spin}}_k(B\Gamma) =: \mathrm{R}^{\mathrm{spin}}_k(\Gamma)$ which we use to identify all these groups with a group which only depends on $\Gamma = \pi_1(M)$. For details compare [**75**].

The groups in the Stolz exact sequence have direct relevance also for the set of concordance classes due to the following result of Stolz [**75**, Proof of Theorem 5.4].

PROPOSITION 10.3. *There is a free and transitive action of the abelian group* $\mathrm{R}^{\mathrm{spin}}_{n+1}(\Gamma)$ *on* $\mathcal{P}^+(M)$. *The inverse of the corresponding action bijection* $\mathrm{R}^{\mathrm{spin}}_{n+1}(\Gamma) \times \mathcal{P}^+(M) \to \mathcal{P}^+(M) \times \mathcal{P}^+(M); (a, x) \mapsto (a \cdot x, x)$ *is given by the map*

$$\begin{aligned} \mathcal{P}^+(M) \times \mathcal{P}^+(M) &\to \mathrm{R}^{\mathrm{spin}}_{n+1}(\Gamma) \times \mathcal{P}^+(M) \\ ([g_0],[g_1]) &\mapsto ((M \times [0,1] \xrightarrow{pr_M} M, g_0 \sqcup g_1), [g_0]), \end{aligned}$$

here we put the two given metrics g_0, g_1 *on the two ends of the cylinder* $W = M \times [0,1]$.

The Stolz positive scalar curvature exact sequence is related to the theme of the present paper (pairing the Higson-Roe exact sequence with cohomology) via [**65**], where a mapping from the positive scalar curvature exact sequence (10.1) to the Higson-Roe analytic exact sequence is constructed:
(10.2)

$$\begin{array}{ccccccccc} \longrightarrow & \Omega^{\mathrm{spin}}_{n+1}(M) & \longrightarrow & \mathrm{R}^{\mathrm{spin}}_{n+1}(\Gamma) & \xrightarrow{\partial} & \mathrm{Pos}^{\mathrm{spin}}_n(M) & \longrightarrow & \Omega^{\mathrm{spin}}_n(M) & \longrightarrow \\ & \downarrow{\scriptstyle \beta} & & \downarrow{\scriptstyle \mathrm{Ind}^\Gamma} & & \downarrow{\scriptstyle \varrho} & & \downarrow{\scriptstyle \beta} & \\ \longrightarrow & K_{n+1}(M) & \xrightarrow{\mu^\Gamma_M} & K_{n+1}(C^*_{red}\Gamma) & \xrightarrow{s} & \mathrm{S}^\Gamma_n(\widetilde{M}) & \xrightarrow{c} & K_n(M) & \xrightarrow{\mu^\Gamma_M} \end{array}$$

This results was established in [**65**] in the even dimensional case (the odd dimensional case was later treated by suspension by Zenobi in [**88**]). Subsequent contributions of Xie and Yu and of Zeidler, see [**82**] [**86**], gave alternative treatments of the result of Piazza and Schick, valid in every dimension. In (10.2) β is the well-known map of generalized homology theory, Ind^Γ is a higher APS-index map, while ϱ associates to a cycle $(f\colon X \to M, g)$ the push-forward of the rho class associated to $\widetilde{D}_g$ on $f^*\widetilde{M}$.

The general pattern to study metrics of positive scalar curvature is now the following: a metric g of positive scalar curvature on M defines a class in $\mathrm{Pos}^{\mathrm{spin}}_n(M)$, which is independent of the concordance class of M, thus there is a well-defined map from $\mathcal{P}^+(M)$ to $\mathrm{Pos}^{\mathrm{spin}}_n(M)$ which associates to the concordance class $[g]$ the bordism class $[M \xrightarrow{\mathrm{id}} M, g]$. Therefore, the following proposition is relevant (which is consistent with the notation of (10.2)). Recall that the compatibility result of [**89**, Section 5.3] gives the following result.

PROPOSITION 10.4. *For a metric* g *of positive scalar curvature on* M *the definition in* (10.2) *of* $\varrho(g) := \varrho([M \xrightarrow{\mathrm{id}} M, g])$ *as given in* [**65**] *coincides with the one we give here in Definition* 3.4*:*

$$\varrho(g) = \varrho(\tilde{D}_g) \in \mathrm{S}^\Gamma_n(\widetilde{M}),$$

as the class of the Dirac operator $\tilde{D}$ *on the universal covering of* M, *through the identification of* $\mathrm{S}^\Gamma_n(\widetilde{M})$ *with the relative K-theory group* $K_{n-1}(C(M) \to \Psi^0_\Gamma(\widetilde{M}))$.

As a consequence, the higher rho numbers defined on $\mathcal{S}^{\Gamma}_n(\widetilde{M})$ can be applied to $\varrho(g)$ and can be used to distinguish elements in $\mathcal{P}^+(M)$ and in particular in $\pi_0(\mathrm{Riem}^+(M))$.

Typically, it is impossible to explicitly compute $\varrho(g)$ or the resulting higher rho numbers. Instead, one uses commutativity of the diagram (10.2) and changes g by elements of $\mathrm{R}^{\mathrm{spin}}_{n+1}(\Gamma)$ and controls how the rho class and the higher rho numbers *change*. Classes in $\mathrm{R}^{\mathrm{spin}}_{n+1}(\Gamma)$, in turn, often have topological origin. For example, we have a factorization $\Omega^{\mathrm{spin}}_{n+1}(M) \to \Omega^{\mathrm{spin}}_{n+1}(B\Gamma) \to \mathrm{R}^{\mathrm{spin}}_{n+1}(\Gamma)$ so that knowledge about $\Omega^{\mathrm{spin}}_{n+1}(B\Gamma)$ can be used to obtain non-trivial elements in $\mathrm{R}^{\mathrm{spin}}_{n+1}(\Gamma)$ and eventually many different classes of metrics of positive scalar curvature on M. This route is exploited e.g. in [**72**]. Equivariant generalized homology classes are used in [**84**] to construct a different type of classes in $\mathrm{R}^{\mathrm{spin}}_{n+1}(\Gamma)$. In both cases, their non-triviality is detected directly in the K-groups of the Higson-Roe analytic exact sequence. We will see that these classes pair non-trivially with our higher cocycles and therefore provide examples of interesting higher rho numbers. A main advantage of these higher rho numbers is the following: in favorable situations our numeric rho invariants obtained from delocalized cyclic cocycles of the fundamental group of a manifold M or from the relative cohomology of $M \to B\Gamma$ can be used not only to distinguish path components of the space of metrics of positive scalar curvature (or concordance classes), but also classes in the moduli space quotient by the action of the diffeomorphism group. It is precisely at this point that the use of these numeric rho invariants is particularly powerful: for these purposes it is much harder to work directly with the rho classes in the analytic structure group $\mathrm{S}^{\Gamma}_n(\widetilde{M})$.

Philosophically, a reason why this has better chances to work is that the large K-theory group $\mathrm{S}^{\Gamma}_n(\widetilde{M})$ is in a certain sense stratified into pieces of different homological nature and degree, and we know a priori that the induced action of the diffeomorphism group will preserve these pieces.

CHAPTER 11

Naturality for the action by the diffeomorphism group

It is well known and already established by Higson and Roe, that the Higson-Roe analytic exact sequence is natural: it can be defined generally for Γ-coverings $\tilde{X} \to X$ (where X is a compact metrizable space), and it is natural for maps of Γ-coverings $(\tilde{X} \to X) \xrightarrow{f} (\tilde{Y} \to Y)$. In this paper, we deal entirely with manifolds $X = M$ and describe the Higson-Roe exact sequence based on algebras of pseudo-differential operators, developed and shown to be equivalent to the original one in [**89**].

In this section, we describe, as special case of naturality, the action of the diffeomorphism group of M in our (pseudodifferential) model of the Higson-Roe exact sequence, and also on the (non-commutative) homology and cohomology groups we are using. We finally show that our maps to homology, respectively our pairings with cohomology, are compatible with the actions.

Let us fix a connected spin manifold M of dimension n with fundamental group $\pi_1(M, m) \cong \Gamma$. To be precise, if we denote by Γ exactly $\pi_1(B\Gamma, *)$, then the previous isomorphism, which will be denoted by $\pi_1(u)$, is induced by the map of pointed spaces $u\colon (M, m) \to (B\Gamma, *)$ which classifies the universal covering $\widetilde{M} \to M$ associated to the basepoint $m \in M$.

Throughout, we assume that M is path-connected.

11.1. Actions of diffeomorphisms

Let us denote by $\mathrm{Diffeo}(M)$ the group of self-diffeomorphisms of M. In this section we are going to describe how this group acts on the K-theoretical and (co)homological objects attached to M.

11.1.1. Action on the fundamental group. Fix an element $\psi \in \mathrm{Diffeo}(M)$. The diffeomorphism ψ can be seen as a map of pointed spaces, hence it induces a group isomorphism $\pi_1(\psi)\colon \pi_1(M, \psi^{-1}(m)) \to \pi_1(M, m)$. Consequently, we observe that only the subgroup of $\mathrm{Diffeo}(M)$ which fixes the basepoint m really acts on $\pi_1(M, m) = \Gamma$.

However, choosing a path γ from $\psi(m)$ to m and concatenating loops with this path and its inverse gives a standard isomorphism $\pi_1(M, m) \to \pi_1(M, \psi^{-1}(m))$. Of course, this isomorphism is non-canonical: it depends on the choice of γ. If we choose a second path γ' we obtain another isomorphism. Their "difference", i.e. the composition of the first such isomorphism with the inverse of the second, is given by conjugation with the loop obtained by concatenating γ with the inverse of γ', i.e. is given by an inner automorphism of $\pi_1(M, m)$.

Therefore, we can identify $\pi_1(M, \psi^{-1}(m))$ with $\pi_1(M, m)$ with an isomorphism which is determined up to post-composition with an inner automorphism of $\pi_1(M, m)$. After this identification, $\pi_1(\psi)$ becomes an automorphism of $\pi_1(M)$, but only well defined up to precomposition with an inner automorphism. This defines a map

$$\alpha\colon \operatorname{Diffeo}(M) \to \operatorname{Out}(\pi_1(M)) = \operatorname{Aut}(\pi_1(M, m))/\operatorname{Inn}(\pi_1(M, m)) \tag{11.1}$$

which is a group homomorphism. Note that, as a consequence of what has just been explained, the outer automorphism groups of $\pi_1(M, m)$ and $\pi_1(M, m_2)$ for two basepoints $m, m_2 \in M$ are canonically isomorphic and therefore it makes sense to write $\operatorname{Out}(\pi_1(M))$, dropping the basepoint.

11.1.2. Action on the universal covering. Let the Γ-covering $\widetilde{M} \to M$ be a universal covering. We can look at the set of all possible lifts of diffeomorphisms of M. This defines a new group $\widetilde{\operatorname{Diffeo}(M)}$ acting by diffeomorphisms on $\widetilde{M}$ and fitting into a group extension

$$1 \to \Gamma \to \widetilde{\operatorname{Diffeo}(M)} \to \operatorname{Diffeo}(M) \to 1. \tag{11.2}$$

The second map assigns to a lifted diffeomorphism the map it covers, and by definition the deck transformation group, identified with Γ via the principal action of Γ on $\widetilde{M}$, is the kernel of this map: the transformations which cover the identity. The map $\widetilde{\operatorname{Diffeo}(M)} \to \operatorname{Diffeo}(M)$ indeed is surjective by the universal lifting property of coverings, the lifting condition is automatically satisfied because the fundamental group of $\widetilde{M}$ is trivial.

REMARK 11.1. Note that, in general, this exact sequence does not split, i.e. we do not have an action of $\operatorname{Diffeo}(M)$ on $\widetilde{M}$. An easy example of the phenomenon is provided by the circle $M = U(1)$: if we look at the subgroup of diffeomorphisms given by $U(1)$ itself (acting by group multiplication, i.e. by rotations by a fixed angle), then the resulting group extension is $0 \to \mathbb{Z} \to \mathbb{R} \to U(1) \to 0$, which does not split.

Let us fix a diffeomorphism $\psi\colon M \to M$ and choose a lift

$$F\colon \widetilde{M} \to \widetilde{M}, \tag{11.3}$$

(a diffeomorphism covering ψ). This diffeomorphism is unique up to pre- or post-composition with a deck-transformation, i.e. an element of Γ.

DEFINITION 11.2. The map F defines a map $\alpha^F : \Gamma \to \Gamma$ as follows: for $x \in \widetilde{M}$, observe that the image of the Γ-orbit of x under F is the Γ-orbit of $F(x)$, $F(x \cdot \Gamma) = F(x) \cdot \Gamma$. Consequently, for $\gamma \in \Gamma$ there is a unique $\alpha^F(\gamma) \in \Gamma$ with

$$F(x \cdot \gamma) = F(x) \cdot \alpha^F(\gamma). \tag{11.4}$$

PROPOSITION 11.3. *The map α^F is an automorphism of Γ and its construction is independent of x. The map $F\colon \widetilde{M} \to \widetilde{M}$ is α^F-equivariant, i.e. $F(x\gamma) = F(x)\alpha^F(\gamma)$ for all $x \in \widetilde{M}$, $\gamma \in \Gamma$. Finally, if we change F by pre-composition with the deck transformation $g \in \Gamma$, the corresponding automorphism $\alpha^{F\circ g}$ is obtained from α^F by pre-composition with the inner automorphism of conjugation with g, $\alpha^{F\circ g}(\gamma) = \alpha^F(g^{-1}\gamma g)\ \forall \gamma \in \Gamma$. If we change F by post-composition with g, we have to post-compose α^F correspondingly and get $\alpha^{g\circ F}(\gamma) = g^{-1}\alpha^F(\gamma)g$ for all $\gamma \in \Gamma$.*

PROOF. First, we show that $\alpha^F(\gamma)$ is independent of $x \in \widetilde{M}$. If $y \in \widetilde{M}$, choose a path p connecting x to y, which exists because M is connected. Then $p\gamma$ is a path connecting $x\gamma$ to $y\gamma$, and $F(p\gamma))$ is a path starting at $F(x)\alpha^F(\gamma)$. On the other hand, $F(p)$ is a path connecting $F(x)$ to $F(y)$ and $F(p)\alpha^F(\gamma)$ also is a path starting at $F(x)\alpha^F(\gamma)$. Both paths $F(p)\alpha^F(\gamma)$ and $F(p\gamma)$ have the same image in M and start at the same point in $\widetilde{M}$, therefore they coincide and indeed $F(y)\alpha^F(\gamma) = F(y\gamma)$ as endpoint of this one path.

If now $\gamma_2 \in \Gamma$ is a second element, then

$$F(x) \cdot (\alpha^F(\gamma_2\gamma)) = F(x \cdot \gamma_2\gamma) = F(x \cdot \gamma_2) \cdot \alpha^F(\gamma) = F(x) \cdot \alpha^F(\gamma_2)\alpha^F(\gamma)$$

so α^F is indeed a homomorphism. As F maps the orbit $x\Gamma$ bijectively to $F(x)\Gamma$, α^F is bijective and hence $\alpha^F \in \mathrm{Aut}(\Gamma)$.

The statement that $F\colon \widetilde{M} \to \widetilde{M}$ is α-equivariant is a reformulation of the fact that the definition of $\alpha^F(\gamma)$ is independent of $x \in \widetilde{M}$. Another direct computation shows that $\alpha^{F\circ g}(\gamma) = \alpha^F(g^{-1}\gamma g)$ and $\alpha^{g\circ F}(\gamma) = g^{-1}\alpha(\gamma)g$ for all $g, \gamma \in \Gamma$. □

LEMMA 11.4. *The outer automorphism $\alpha(\psi)$ assigned to ψ by* (11.1) *coincides with the outer automorphism induced by α^F of Proposition* 11.3. *Note that, as outer isomorphism, by Proposition* 11.3 *α^F is independent of the particular lift F of ψ.*

PROOF. By definition, the (outer) automorphism α^F is determined by $F(\tilde{x}\gamma) = F(\tilde{x})\alpha^F(\gamma)$ for an arbitrary $\tilde{x} \in \widetilde{M}$ and for $\gamma \in \Gamma$.

Let $\gamma \in \Gamma$ be represented by a loop starting at $x = p(\tilde{x}) \in M$, and lift it to a path $\tilde{\gamma}$ from $\tilde{x}$ to $\tilde{x}\gamma$. Then $F(\tilde{x}\gamma) = F(x) \cdot \alpha^F(\gamma)$ is the endpoint of the path $F(\tilde{\gamma})$ starting at $F(\tilde{x})$. Its image under the covering projection $p\colon \widetilde{M} \to M$ is the image under ψ of the initial loop $p(\tilde{\gamma})$ representing γ, i.e. is $\pi_1(\psi)(\gamma)$, so that $\alpha(\psi)(\gamma) = \pi_1(\psi)(\gamma) = \alpha^F(\gamma)$. Note that, modulo inner automorphisms, we can ignore base points here and everything is well defined, and the claim is proved. □

The following observation is straightforward.

LEMMA 11.5. *Assume we have a vector bundle E over M and a vector bundle isomorphism $\Psi\colon E \to E$ covering the diffeomorphism $\psi\colon M \to M$. Denote by $\widetilde{E}$ the (equivariant) lift of E to $\widetilde{M}$.*

Then Ψ pulls back to a vector bundle isomorphism $\widetilde{\Psi}\colon \widetilde{E} \to \widetilde{E}$ covering every lift $F\colon \widetilde{M} \to \widetilde{M}$ of f.

11.1.3. Action on pseudodifferential operators. It is a standard fact, see for instance Theorem 4.1 in Shubin's book [**73**, Formula 4.2], that if X is a smooth manifold, then Diffeo(X) acts on properly supported pseudodifferential operators by conjugation through pull-back operators. More precisely, let $f \in \mathrm{Diffeo}(X)$ and K a properly supported pseudodifferential operator with distributional kernel $K(x, y)$. Then f_*K is obtained as $(f^{-1})^* \circ K \circ f^*$, with distributional kernel

$$f_*K(x, y) := K(f^{-1}(x), f^{-1}(y)).$$

Now let us concretely describe the action of $f \in \mathrm{Diffeo}(M)$ on the pseudodifferential operator algebras showing up in our description of the Higson-Roe sequence. Let $K(x, y)$ be the kernel of an element K in $\Psi^0_{\Gamma,c}(\widetilde{M})$, i.e. K is a Γ-equivariant distribution on $\widetilde{M} \times \widetilde{M}$ for the diagonal action of Γ. Let $\tilde{f} \in \widetilde{\mathrm{Diffeo}(M)}$ be a lift of

f. Then, since $\widetilde{f}$ enjoys property (11.4), it is immediate that $\widetilde{f}_*K$ is Γ-equivariant and that this action does not depend on the choice of the lift $\widetilde{f}$. Moreover, this action preserves both the ideal of smoothing operators and the subalgebra $C(M)$, where the action reduces to the usual action of $f \in \text{Diffeo}(M)$ on $C(M)$ by pre-composition with f^{-1}. By continuity, this extends to an action of Diffeo(M) by automorphisms of C*-algebras.

Finally, let us consider the case where our pseudodifferential operators act on the sections of a given vector bundle $E \to M$.

DEFINITION 11.6. We now need a slight refinement of (11.2). This is given by a subgroup $\text{Diffeo}_0(M)$ of Diffeo(M) and an extension $1 \to K \to \text{Diffeo}(E) \xrightarrow{\pi} \text{Diffeo}_0(M) \to 1$, together with a lift of the action of $\text{Diffeo}_0(M)$ on M to an action of Diffeo(E) on E through bundle automorphisms such that $\psi \in \text{Diffeo}(E)$ covers $\pi(\psi) \in \text{Diffeo}(M)$. Then all the constructions work in a same way and give us an action of Diffeo(E) by automorphisms of C*-algebras. Notationally, we will ignore these details and write Diffeo(M) throughout, but we should have in mind that, strictly speaking, it has to be replaced by Diffeo(E) where appropriate.

REMARK 11.7. The situation we have in mind here is the one where E is the spinor bundle of a spin structure. Then $\text{Diffeo}_0(M)$ is the subgroup of Diffeo(M) of diffeomorphisms which are covered by a map of the spin principal bundle. There are always two such lifts, and Diffeo(E) constitutes the resulting bundle maps, a 2-fold covering of $\text{Diffeo}_0(M)$. In this case, $K = \mathbb{Z}/2$.

11.1.4. Action on the Higson-Roe exact sequence. Let M be a closed manifold and E a vector bundle over M such that we are in the situation of Definition 11.6. Then, by what we have just seen, Diffeo(E) naturally acts on the Higson-Roe exact sequence

$$\xrightarrow{\partial} K_*(0 \to \Psi^{-\infty}_{\mathcal{A}\Gamma}(\widetilde{M},\widetilde{E})) \xrightarrow{i_*} K_*(C^\infty(M) \xrightarrow{\mathrm{m}} \Psi^0_{\mathcal{A}\Gamma}(\widetilde{M},\widetilde{E}))$$
$$\xrightarrow{\sigma_*} K_*(C^\infty(M) \to \Psi^0_{\mathcal{A}\Gamma}(\widetilde{M},\widetilde{E})/\Psi^{-\infty}_{\mathcal{A}\Gamma}(\widetilde{M},\widetilde{E})) \xrightarrow{\partial}$$

11.1.5. Action on the Stolz positive scalar curvature exact sequence. The group Diffeo(M), or more generally any subgroup U of the homeomorphism group of M acts on the whole Stolz exact sequence, simply by post-composition of a cycle, which is represented by some object together with a map to M, with the homeomorphism. It is clear that this is compatible with the maps in the Stolz exact sequence. The canonical isomorphism $\mathrm{R}^{\mathrm{spin}}_*(M) \to \mathrm{R}^{\mathrm{spin}}_*(B\Gamma)$ implies that the action on this term of the Stolz exact sequences factors through $\text{Out}(\pi_1(M))$.

If a cycle is represented by a metric g on M itself, we have a competing option to let $\psi \in \text{Diffeo}(M)$ act, namely by pulling back g with ψ. Fortunately, these two options coincide:

PROPOSITION 11.8. *Let $(f\colon M \to M, g)$ represent a class in $\mathrm{Pos}^{\mathrm{spin}}_n(M)$ and let ψ be a spin structure preserving diffeomorphism of M. Then*

$$[(\psi \circ f\colon M \to M, g)] = [(f\colon M \to M, \psi_* g)] \in \mathrm{Pos}^{\mathrm{spin}}_n(M).$$

Similarly, let $(f\colon M \times [0,1] \to M, g_0 \sqcup g_1)$ with g_0, g_1 Riemannian metrics of positive scalar curvature on $M \times \{0\}$ and $M \times \{1\}$ represent a class in $\mathrm{R}^{\mathrm{spin}}_{n+1}(M)$. Then

$$[(\psi \circ f\colon M \times [0,1] \to M, g_0 \sqcup g_1)] = [(f\colon M \times [0,1] \to M, \psi_* g_0 \sqcup \psi_* g_1)] \in \mathrm{R}^{\mathrm{spin}}_{n+1}(M).$$

PROOF. We construct the spin bordism $W = M\times[0,1/2]\cup_\psi M\times[1/2,1]$ where we glue x in the left copy to $\psi(x)$ in the right copy. Note that W inherits a spin structure from the action of ψ on the spin principal bundle which is given because we assume that ψ is spin structure preserving. Then $f\circ\psi\times \mathrm{id}_{[0,1/2]}\cup f_M\times \mathrm{id}_{[1/2,1]}$ is a well defined continuous map $W\to M$. Moreover, the metric $g+dt^2$ on $M\times[0,1/2]$ glues isometrically to ψ_*g+dt^2 on $M\times[1/2,1]$, defining a metric g_W of positive scalar curvature on W. We obtain the desired bordism between $(\psi\circ f\colon M\to M, g)$ and $(f\colon M\to M,\psi_*g)$.

The proof for $\mathrm{R}^{\mathrm{spin}}_{n+1}(M)$ is identical. □

COROLLARY 11.9. *The free and transitive action of* $\mathrm{R}^{\mathrm{spin}}_{n+1}(M)$ *on* $\mathcal{P}^+(M)$ *is compatible with the action of* $\mathrm{Diffeo}(M)$ *on both sets.*

PROOF. This is a direct consequence of the defining formula in Proposition 10.3 and Proposition 11.8. □

11.1.6. Action on the cyclic (co)homology of group algebras. Recall that we have a well defined homomorphism (independent of issues with basepoints) $\alpha\colon \mathrm{Diffeo}(M)\to \mathrm{Out}(\pi_1(M))$.

Fortunately, the invariants we care about are of (co)homological nature for the fundamental group, and inner automorphisms act trivially on (co)homology. Moreover, the (co)homological constructions we apply are clearly natural. In particular, we have the following consequence:

PROPOSITION 11.10. *There are well defined natural actions of* $\mathrm{Out}(\Gamma)$ *and via the homomorphism* α *to* $\mathrm{Out}(\Gamma)$ *also of* $\mathrm{Diffeo}(M)$ *on*

$$H_*(\mathcal{A}\Gamma)\hookrightarrow HC_*(\mathcal{A}\Gamma)\to HP_*(\mathcal{A}\Gamma)$$

and dually on

$$HC^*(\mathbb{C}\Gamma)\to HP^*(\mathbb{C}\Gamma)\to HP^*(\mathcal{A}\Gamma)\to HP_*(\mathcal{A}\Gamma)'\to H_*(\mathcal{A}\Gamma)'$$

here $'$ *denotes the dual space, and cohomology maps to the dual space of homology via the usual pairing of cohomology and homology.*

The corresponding statements hold for localized and delocalized (co)homology like $H^{del}_*(\mathcal{A}\Gamma)$.

PROOF. By naturality of the constructions, it is evident that the true automorphism group $\mathrm{Aut}(\Gamma)$ acts on the whole sequence and the maps are equivariant.

Moreover, it is a standard fact that inner automorphisms induce the trivial map on group (co)homology. For the (co)homology groups relevant here this already holds on the level of the defining cochain complexes. Recall that they are obtained from the universal DGA over $\mathcal{A}\Gamma$ taking the quotient by the subspace generated by commutators. For an inner automorphism $\alpha(\gamma)=g\gamma g^{-1}$, and a non-commutative form ω we directly have that $\alpha(\omega)-\omega$ is a commutator. For example, $g\gamma g^{-1}-\gamma = g(\gamma g^{-1})-(\gamma g^{-1})g$ is the commutator of g and γg^{-1}. Similarly, $d(g\gamma g^{-1})-d\gamma = dg\cdot(\gamma g^{-1})+gd(\gamma g^{-1})-d(\gamma g^{-1})\cdot g-(\gamma g^{-1})dg$ is the sum of the commutators of g and $d(\gamma g^{-1})$ and of dg and (γg^{-1}). The proof for general ω is left to the reader. □

11.1.7. Naturality of the Chern character. We now show that the action of $\mathrm{Diffeo}(M)$ (or rather $\mathrm{Diffeo}(E)$ in the case of the additional data of a bundle $E\to M$ as in Definition 11.6) extends to a canonical and compatible action on the whole Chern character sequence (6.19), the main construction of this paper. In fact, the action on the homological part is given through the homomorphism $\mathrm{Diffeo}(E) \to \mathrm{Diffeo}(M) \to \mathrm{Out}(\Gamma)$ and is the canonical action of the outer automorphism group on group (co)homology.

More specifically, we will prove the following result.

THEOREM 11.11. *Let M be connected with fundamental group Γ, $\widetilde{M} \to M$ a universal covering, considered as Γ-principal bundle. Fix $\psi \in \mathrm{Diffeo}(M)$. Fix an additional bundle $E \to M$ and assume we haven an extension $\mathrm{Diffeo}(E) \to \mathrm{Diffeo}_0(M) \to 1$ as in Definition 11.6.*

For every $\psi \in \mathrm{Diffeo}(E)$ we obtain a commutative diagram

$$\begin{array}{ccc} K_*(C^\infty(M) \xrightarrow{\mathrm{m}} \Psi^0_{\mathcal{A}\Gamma}(\widetilde{M}, \widetilde{E})) & \xrightarrow[\cong]{\psi_*} & K_*(C^\infty(M) \xrightarrow{\mathrm{m}} \Psi^0_{\mathcal{A}\Gamma}(\widetilde{M}, \widetilde{E})) \\ \downarrow{\scriptstyle \mathrm{Ch}^{del}_\Gamma} & & \downarrow{\scriptstyle \mathrm{Ch}^{del}_\Gamma} \\ H^{del}_{[*-1]}(\mathcal{A}\Gamma) & \xrightarrow[\cong]{\alpha(\psi)_*} & H^{del}_{[*-1]}(\mathcal{A}\Gamma) \end{array}$$

and correspondingly for the other terms of the sequence (6.19), compatible with the maps in (6.19).

The whole construction is natural in the diffeomorphism ψ, i.e. we obtain an natural action of $\mathrm{Diffeo}(M)$ *on the diagram (6.19), where the action on cohomology is through the homomorphism* $\alpha\colon \mathrm{Diffeo}(E) \to \mathrm{Diffeo}(M) \to \mathrm{Out}(\Gamma)$.

We claim that, to a large extent, the assertion of Theorem 11.11 is tautological, given that our constructions in Sections 5 and 6 are natural and canonical.

To observe this, we formulate two technical propositions which displays this fact.

PROPOSITION 11.12. *Let $\widetilde{M} \to M$ be a Γ-principal bundles over a smooth manifold M, and $\tilde{N} \to N$ a Γ_2-principal bundle over the smooth manifold N. Let $\alpha\colon \Gamma_2 \to \Gamma$ be a group isomorphism, $f\colon N \to M$ be a diffeomorphism, covered by an α-equivariant diffeomorphism $F\colon\ N \to \widetilde{M}$, i.e. $F(x\cdot\gamma) = F(x)\cdot\alpha(\gamma)$ for all $x \in \tilde{N}$, $\gamma \in \Gamma_2$. Let $E_M \to M$ and $E_N \to N$ be vector bundles with a bundle isomorphism $\Psi\colon E_N \to E_M$ covering the diffeomorphism $\psi\colon N \to M$.*

Then the maps (F, Ψ, α) induce tautological isomorphisms making the following diagram commutative

$$\begin{array}{ccccc} \cdots \longrightarrow & K_*(0 \to \Psi^{-\infty}_{\mathcal{A}\Gamma_2}(\widetilde{N}, \widetilde{E}_N)) & \xrightarrow{i_*} & K_*(C^\infty(N) \xrightarrow{\mathrm{m}} \Psi^0_{\mathcal{A}\Gamma_2}(\widetilde{N}, \widetilde{E}_N)) \\ & \cong\downarrow{\scriptstyle (F,\Psi,\alpha)_*} & & \cong\downarrow{\scriptstyle (F,\Psi,\alpha)_*} \\ \cdots \longrightarrow & K_*(0 \to \Psi^{-\infty}_{\mathcal{A}\Gamma}(\widetilde{M}, \widetilde{E}_M)) & \xrightarrow{i_*} & K_*(C^\infty(M) \xrightarrow{\mathrm{m}} \Psi^0_{\mathcal{A}\Gamma}(\widetilde{M}, \widetilde{E}_M)) \end{array}$$

$$\begin{array}{ccc} K_*(C^\infty(N) \xrightarrow{\mathrm{m}} \Psi^0_{\mathcal{A}\Gamma_2}(\widetilde{N}, \widetilde{E}_N)) & \xrightarrow{\sigma_*} & K_*(C^\infty(N) \xrightarrow{\pi^*} \Psi^0_{\mathcal{A}\Gamma_2}(\widetilde{N}, \widetilde{E}_N)/\Psi^{-\infty}_{\mathcal{A}\Gamma_2}(\widetilde{N}, \widetilde{E}_N)) \\ \cong\downarrow{\scriptstyle (F,\Psi,\alpha)_*} & & \cong\downarrow{\scriptstyle (F,\Psi,\alpha)_*} \\ K_*(C^\infty(M) \xrightarrow{\mathrm{m}} \Psi^0_{\mathcal{A}\Gamma}(\widetilde{M}, \widetilde{E}_M)) & \xrightarrow{\sigma_*} & K_*(C^\infty(M) \xrightarrow{\pi^*} \Psi^0_{\mathcal{A}\Gamma}(\widetilde{M}, \widetilde{E}_M)/\Psi^{-\infty}_{\mathcal{A}\Gamma}(\widetilde{M}, \widetilde{E}_M)) \end{array}$$

All the vertical isomorphism are compatible with the Chern character maps of Theorem 6.20 if we map the group homology of Γ_2 to the one of Γ via the map induced *by the group isomorphism α. Explicitly, the diagram*

$$\begin{array}{ccc} K_*(C^\infty(N) \xrightarrow{\mathrm{m}} \Psi^0_{\mathcal{A}\Gamma_2}(\widetilde{N}, \widetilde{E}_N)) & \xrightarrow[\cong]{(F,\psi,\alpha)_*} & K_*(C^\infty(M) \xrightarrow{\mathrm{m}} \Psi^0_{\mathcal{A}\Gamma}(\widetilde{M}, \widetilde{E}_M)) \\ \downarrow \mathrm{Ch}^{del}_{\Gamma_2} & & \downarrow \mathrm{Ch}^{del}_{\Gamma} \\ H^{del}_{[*-1]}(\mathcal{A}\Gamma_2) & \xrightarrow[\cong]{\alpha_*} & H^{del}_{[*-1]}(\mathcal{A}\Gamma) \end{array}$$

is commutative, as well as the corresponding other diagrams obtained from the sequence (6.19). *Moreover, the tautological isomorphisms we are using here are natural (in the triples (F, ψ, α)).*

PROOF. The effect of the maps α f, F, Ψ is the one of renaming the elements of Γ and the points of $M, \widetilde{M}, E$, preserving all bits of the structure which is used in the constructions of Sections 5 and 6. This implies that the maps induce natural isomorphisms of the algebras and DGAs constructed, compatible with all the maps we apply. The statement for the induced K-theory and the compatibility with the Chern character follows, and all of this is tautological. □

PROOF OF THEOREM 11.11. The proof of the main result of this section, the fact that we have an action of the diffeomorphism group on the whole diagram (6.19) (the Chern character map from the Higson-Roe analytic sequence to group homology) follows directly from the abstract compatibility Proposition 11.12.

Let us discuss the details of this derivation: It is a direct consequence of Proposition 11.12 and Proposition 11.3 that we have an action of $\widetilde{\mathrm{Diffeo}}(E)$, the extension of $\mathrm{Diffeo}(E)$ constructed like $\widetilde{\mathrm{Diffeo}}(M)$ in (11.2), and that the corresponding action on group homology is via $\alpha^F \in \mathrm{Aut}(\Gamma)$ for $F \in \widetilde{\mathrm{Diffeo}}(E)$.

It remains to check that the action factors through $\mathrm{Diffeo}(E)$ and through the image of $\alpha^F \in \mathrm{Out}(\Gamma)$, which by Lemma 11.4 is indeed $\alpha(\psi)$ if F is a lift of $\psi \in \mathrm{Diffeo}(E)$.

The statement about group homology is precisely Proposition 11.10.

For the action on the Higson-Roe exact sequence, we observe that the terms are given as (relative) K-theory groups of algebras of Γ-equivariant pseudodifferential operators on $\widetilde{M}$. We have already discussed in Subsection 11.1.3 that this action, which is by pull-back of the distributional kernel on $\widetilde{M} \times \widetilde{M}$ does not depend on the particular lift F of a diffeomorphism $\psi \in \mathrm{Diffeo}(E)$. The claim follows. □

COROLLARY 11.13. *Assume that M is a connected manifold with universal covering the Γ-principal bundle $\widetilde{M} \to M$. Fix $\psi \in \mathrm{Diffeo}(M)$ with $\alpha = \pi_1(\psi) \in \mathrm{Out}(\Gamma)$, $x \in \mathrm{S}^\Gamma_*(\widetilde{M})$, and $\tau \in HC^{[*-1]}_{del}(\mathcal{A}\Gamma)$.*

Then we have naturality of the pairing between K-theory and cohomology

$$\langle \psi_* x, \tau \rangle := \langle \mathrm{Ch}^{del}_\Gamma(\psi_* x), \tau \rangle \overset{(a)}{=} \langle \alpha_* \mathrm{Ch}^{del}_\Gamma(x), \tau \rangle \overset{(b)}{=} \langle \mathrm{Ch}^{del}_\Gamma(x), \alpha^* \tau \rangle = \langle x, \alpha^* \tau \rangle.$$

PROOF. We define the pairing between K-theory and cohomology through the Chern character map from K-theory to homology and then the standard pairing between homology and cohomology. Equation (a) is just the naturality assertion of Theorem 11.11. Equation (b) is the naturality of the pairing between homology and cohomology, here for the algebra automorphism induced by α (or rather for any automorphism of Γ representing the outer automorphism α), as detailed in Proposition 11.10. □

REMARK 11.14. In Section 7.6 we give rather explicit integral formulae for the pairings in Corollary 11.13. From these explicit formulae, one can then also see directly that the pairing is natural as stated in Corollary 11.13. This constitutes an alternative proof of Corollary 11.13 which also implies Theorem 11.11 due to the fact that the pairing between homology and cohomology is sufficiently non-degenerate.

PROPOSITION 11.15. *Let g be a metric with positive scalar curvature on M and $\psi \in \mathrm{Diffeo}(M)$ with a lift to the spin principle bundle (i.e. ψ "is spin structure preserving") then*

$$\psi_*(\varrho(\psi^* g)) = \varrho(g) \in K_*(C(M) \to \Psi^0_\Gamma(\widetilde{M})). \tag{11.5}$$

PROOF. Let $\tilde{\psi}$ be a lift of ψ to $\widetilde{M}$. As in the proof of [**64**, Proposition 2.10] one can prove that $\tilde{D}_{\psi^* g}$ is obtained from $\tilde{D}_g$ by unitary conjugation with $\tilde{\psi}^*$, or in other words $\tilde{\psi}_*(\tilde{D}_{\psi^* g}) = \tilde{D}_g$. By naturality of functional calculus, this is also true for the sign (and other functions) of these operators. Then, by Definition 3.4, $\varrho(\psi^* g)$ is an explicitly given function of $\mathrm{sgn}(D_{\psi^* g})$ and hence (11.5) follows. □

COROLLARY 11.16. *Let g be a metric with positive scalar curvature on M, $\psi \in \mathrm{Diffeo}(M)$ with a lift to the spin principle bundle and $\tau \in HC^{[*-1]}_{del}(\mathcal{A}\Gamma)$, then*

$$\langle \varrho(g), \tau \rangle = \langle \varrho(\psi^* g), \alpha^* \tau \rangle \tag{11.6}$$

with $\alpha = \pi_1(\psi) \in \mathrm{Out}(\Gamma)$.

11.1.8. Action on the long exact sequence in cohomology of the pair $(B\Gamma, M)$. The action of $\psi \in \mathrm{Diffeo}(M)$ on cocycles is obviously given by pull-back. Moreover it induces a natural automorphism of the whole long exact sequence of cohomology groups.

(11.7)

$$\begin{array}{ccccccccc}
\longrightarrow & H^{*-1}(M) & \xrightarrow{\delta} & H^*(M \to B\Gamma) & \longrightarrow & H^*(B\Gamma) & \xrightarrow{u^*} & H^*(M) & \longrightarrow \\
& \downarrow{\scriptstyle \psi^*} & & \downarrow{\scriptstyle (\psi, B\alpha_\psi)^*} & & \downarrow{\scriptstyle B\alpha_\psi^*} & & \downarrow{\scriptstyle \psi^*} & \\
\longrightarrow & H^{*-1}(M) & \xrightarrow{\delta} & H^*(M \to B\Gamma) & \longrightarrow & H^*(B\Gamma) & \xrightarrow{u^*} & H^*(M) & \longrightarrow
\end{array}$$

This works also in the model with Alexander-Spanier cochains on $\widetilde{M}$ in Section 8.1, via a lift $\tilde{\psi}$ of ψ to $\widetilde{M}$. As before, the action on cohomology is independent of the lift. In particular, no issues with base points (in the cohomology of $B\Gamma$) occur. With little effort, it follows that the pairings of Theorem 8.24, Definition 8.26 and the whole diagram (8.19) are compatible with the action of $\mathrm{Diffeo}(M)$.

11.2. Moduli spaces of metrics of positive scalar curvature

Given a smooth manifold M, the space of metrics of positive scalar curvature arguably is not the most relevant object: should the many different but isometric Riemannian metrics really be all distinguished? This is why we are interested in the *moduli space of Riemannian metrics of positive scalar curvature* on M

$$\mathrm{Riem}^+(M)/\,\mathrm{Diffeo}(M),$$

where we divide by the action of the diffeomorphism group, a diffeomorphism mapping one metric to the other being an isometry between the two metrics.

However, in geometry there are important metric/topological concepts which are not preserved by this concept. One of these, studied intensely in hyperbolic geometry, is the marked length spectrum.

DEFINITION 11.17. Let (M, g) be a closed Riemannian manifold. Its *marked length spectrum* is the map which assigns to each free homotopy class of closed loops (free means that the loops don't have to respect a basepoint) the length of the shortest closed geodesic in its class.

Less information contains the **unmarked length spectrum**, which just is the subset of the real line consisting of the numbers we obtain this way (listed with multiplicity).

In light of this important invariant, we propose to study not only the usual moduli space, but also the *marked moduli space*. This is obtained as the quotient by the *marked diffeomorphism group*

$$\operatorname{Diffeo}_m(M) := \{\phi \in \operatorname{Diffeo}(M) \mid \phi_* = 1 \in \operatorname{Out}(\pi_1(M))\},$$

those diffeomorphisms whose induced outer automorphism of the fundamental group is trivial.

Note that the set of free homotopy classes of loops in M is canonically in bijection to the set of conjugacy classes in $\pi_1(M)$ (and the latter set does not depend on the basepoint). This means that the marked diffeomorphisms are precisely those which act trivially on the set of free homotopy classes of loops. As a consequence, a *marked isometry* between two metrics g_1, g_2 on M, i.e. an isometry induced by a marked diffeomorphisms, preserves the marked length spectrum.

We therefore propose to study besides the usual moduli space also the following object.

DEFINITION 11.18. We define the *marked moduli space* of metrics of positive scalar curvature on M as the space $\operatorname{Riem}^+(M)/\operatorname{Diffeo}_m(M)$.

PROPOSITION 11.19. *Let M be a connected spin manifold of dimension n. Assume that (M, g_1) and (M, g_2) are two metrics of positive scalar curvature which represent two different classes in $\operatorname{Pos}_n^{\mathrm{spin}}(X)$ for a fixed reference map $u\colon M \to X$.*

Then there is no diffeomorphism $\phi\colon M \to M$ such that $u \circ \phi$ is homotopic to u and such that $g_2 = \phi^* g_1$. In the special case that $X = B\pi_1(M)$ and $u\colon M \to B\pi_1(M)$ induces the identity map of fundamental groups, the condition $u \circ \phi \simeq u$ means precisely that ϕ induces the identity outer automorphism of $\pi_1(M)$.

PROOF. Given a diffeomorphism ϕ with $\phi^* g_1 = g_2$ and $u \simeq u \circ \phi$ we can produce a psc-bordism $W \to B\Gamma$ as follows: $W = M \times [0, 1/2] \cup_\phi M \times [1/2, 1]$, i.e. we glue to cylinders over M using the diffeomorphism ϕ. On the first half, we put the metric $g_1 + dt^2$ and on the second half we put $g_2 + dt^2$. The fact that ϕ is an isometry between g_1 and g_2 means that this is a well defined smooth metric on W. As reference map to $B\Gamma$ we use $u \circ pr_M$ on the first half of the cylinder, and the homotopy between $u \circ \phi$ and u on the second half of the cylinder.

It is a standard fact that two maps to $B\Gamma$ are homotopic relative to a basepoint if and only if they induce the same map on fundamental groups, and two maps are homotopic not necessarily preserving the basepoint if and only if they induce the same map up to an inner automorphism. As $u_* = \mathrm{id}_{\pi_1(M)}$ this is satisfied for the maps u and $u \circ \phi$ precisely if ϕ induces the trivial outer automorphism of $\pi_1(M)$. $\square$

The following theorem is a direct corollary of Proposition 11.19.

THEOREM 11.20. *Assume that we are in the situation of Proposition 11.19: we have metrics g_1 and g_2 of positive scalar curvature in M which represent different elements in $\mathrm{Pos}_n^{\mathrm{spin}}(B\pi_1(M))$. Then g_1 and g_2 lie in different components of the marked moduli space $\mathrm{Riem}^+(M)/\mathrm{Diffeo}_m(M)$.*

In other words: non-triviality of $\mathrm{Pos}_n^{\mathrm{spin}}(B\Gamma)$ immediately gives non-triviality of the set of components of the marked moduli space $\pi_0(\mathrm{Riem}^+(M)/\mathrm{Diffeo}_m(M))$.

REMARK 11.21. It is much more challenging to prove in the context of Theorem 11.20 that the metrics also lie in different components of the full moduli space $\mathrm{Riem}^+(M)/\mathrm{Diffeo}(M)$. Here, our higher rho numbers can be used.

11.3. Defining the size of the moduli space

We are interested in measuring the "size" of the moduli space of metrics of structure. More specifically, recall that we have a surjective map $\pi_0(\mathrm{Riem}^+(M)) \to \mathcal{P}^+(M)$, where $\mathcal{P}^+(M)$ is the set of concordance classes of metrics of positive scalar curvature on M.

Next, as we have already seen in Proposition 10.3, there is a free and transitive action of the abelian group $\mathrm{R}_{n+1}^{\mathrm{spin}}(M)$ on $\mathcal{P}^+(M)$ (if M is a spin manifold of dimension $n \geq 5$) with fundamental group Γ, compare the proof of [**75**, Theorem 5.4]. Occasionally, one defines a (non-canonical) group structure on $\mathcal{P}^+(M)$ by choosing a basepoint $g_0 \in \mathcal{P}^+(M)$ and then identifying $\mathcal{P}^+(M)$ with the group $\mathrm{R}_{n+1}^{\mathrm{spin}}(M)$ via the action of $\mathrm{R}_{n+1}^{\mathrm{spin}}(M)$ on g_0. In any event, it is customary to measure the size of $\mathcal{P}^+(M)$ as the rank of the group $\mathrm{R}_{n+1}^{\mathrm{spin}}(M)$ which acts freely and transitively on it.

Because $\pi_0(\mathrm{Riem}^+(M))$ surjects onto $\mathcal{P}^+(M)$, we can then say that the "rank of $\pi_0(\mathrm{Riem}^+(M))$" in this sense is bounded below by the rank of $\mathrm{R}_{n+1}^{\mathrm{spin}}(M)$, even though $\pi_0(\mathrm{Riem}^+(M))$ in general has no abelian group structure.

Apart from the concordance relation, we often use the bordism relation for metric of positive scalar curvature. It is now a bit more tricky to define ranks for the moduli space of metrics of positive scalar curvature. The standard way how this is done, interpreted slightly differently from the literature, is discussed in the following.

We have the canonically defined map

$$\pi_0(\mathrm{Riem}^+(M)) \to \mathcal{P}^+(M) \to \mathrm{Pos}_n^{\mathrm{spin}}(M); [g] \mapsto (M, g, \mathrm{id}),$$

where a metric of positive scalar curvature is mapped to the bordism class of $\mathrm{id}\colon M \to M$, with metric g on the source M. Note that the image of $\mathcal{P}^+(M)$ is not a subgroup, but a coset of the image of $\partial\colon \mathrm{R}_{n+1}^{\mathrm{spin}}(\Gamma) \to \mathrm{Pos}_n^{\mathrm{spin}}(M)$; indeed, it is the inverse image of the fundamental class $[M \xrightarrow{\mathrm{id}} M] \in \Omega_n^{\mathrm{spin}}(M)$ under the map $\mathrm{Pos}_n^{\mathrm{spin}}(M) \to \Omega_n^{\mathrm{spin}}(M)$ of the Stolz exact sequence (10.2) and thus a coset of the kernel of this map.

DEFINITION 11.22. We define the set $\Omega\mathcal{P}^+(M)$ of bordism classes of metrics of positive scalar curvature on M as

$$\Omega\mathcal{P}^+(M) := \mathrm{im}(\mathcal{P}^+(M) \to \mathrm{Pos}_n^{\mathrm{spin}}(M)).$$

Concretely, this means that we look at the space of metrics of positive scalar curvature on M but impose the even stronger spin bordism relation instead of the concordance relation.

We define the *rank of the set of bordism classes of metrics of positive scalar curvature on M* as the rank of the coset $\mathrm{im}(\mathcal{P}^+(M) \to \mathrm{Pos}_n^{\mathrm{spin}}(M))$; this is, by definition, the rank of the corresponding group $\mathrm{im}(\mathrm{R}_{n+1}^{\mathrm{spin}}(M) \to \mathrm{Pos}_n^{\mathrm{spin}}(M))$.

Now we consider the action of the group $\mathrm{Diffeo}(M)$ on the whole diagram. By Proposition 11.8 and Corollary 11.13, all the maps we consider are equivariant for this action. Our goal is to study the quotients by this action. When it comes to the abelian groups $\mathrm{R}_*^{\mathrm{spin}}(M)$ and $\mathrm{Pos}_*^{\mathrm{spin}}(M)$, the set of orbits will of course not have a group structure. Instead, one passes to the coinvariant group. Recall that if a group G acts on an abelian group A, the coinvariant group A_G is defined as the quotient of A by the subgroup generated by the set $\{a - ag \mid a \in A, g \in G\}$.

In our situation, for any subgroup $U \subset \mathrm{Diffeo}(M)$ we then have for each choice of a basepoint $g_0 \in \mathcal{P}^+(M)$ a surjective map

$$\mathcal{P}^+(M)/U \to \mathrm{R}_{n+1}^+(M)_U$$

induced by the inverse of the action map, and correspondingly a well-defined chain of maps

$$\pi_0(\mathrm{Riem}^+(M)/U) = \pi_0(\mathrm{Riem}^+(M))/U \twoheadrightarrow \mathcal{P}^+(M)/U \to \mathrm{Pos}_n^{\mathrm{spin}}(M)_U$$

where the image in $\mathrm{Pos}_n^{\mathrm{spin}}(M)_U$ is a coset of the image group $\mathrm{R}_{n+1}^{\mathrm{spin}}(M)_U \to \mathrm{Pos}_n^{\mathrm{spin}}(M)_U$.

To measure the "size" of the moduli space of Riemannian metrics of positive scalar curvature on M, or the concordance moduli space, it is now customary to use the rank of the coinvariant groups $\mathrm{R}_{n+1}^{\mathrm{spin}}(M)_U$ or of its image in $\mathrm{Pos}_n^{\mathrm{spin}}(M)_U$, where the latter describes the size of a psc-bordism moduli space.

REMARK 11.23. However, there is one subtlety here: when passing from a group U which acts on an abelian group A to a subgroup U_2 of finite index, the rank of the coinvariant group can change. Take, as an example $A = \mathbb{Z}$ and $U = \{\mathbb{Z}/2\}$ acting by reflection. Then A_U is finite, whereas of course $A_{\{1\}} = A$ so that the rank of the coinvariant group increases from 0 to 1 if we replace the acting group U by the trivial group, which has index 2 in U.

We propose that we get a better idea of the size of the moduli space if we consider all subgroups of finite index in $\mathrm{Diffeo}(M)$, because it might just happen that some elements of $\mathrm{Diffeo}(M)$ have some "folding" action on $\mathrm{R}_{n+1}^{\mathrm{spin}}(M)$ or $\mathrm{Pos}_n^{\mathrm{spin}}(M)$ which artificially reduce the rank of the coinvariant space.

DEFINITION 11.24. The *virtual rank of the concordance moduli space* $\widetilde{\mathcal{P}^+}(M)$ of Riemannian metrics of positive scalar curvature on M is defined as

$$\begin{aligned}\mathrm{vrank}(\widetilde{\mathcal{P}^+}(M)) &:= \mathrm{vrank}(\text{“}\mathcal{P}^+(M)/\,\mathrm{Diffeo}(M)\text{”})\\ &:= \sup\{\mathrm{rank}(\mathrm{R}^{\mathrm{spin}}(M)_U) \mid U \le \mathrm{Diffeo}(M) \text{ of finite index}\}.\end{aligned}$$

Similarly, we define the *virtual rank of the bordism moduli space* $\widetilde{\Omega\mathcal{P}^+}$ of Riemannian metrics of positive scalar curvature on M as

$$\begin{aligned}&\mathrm{vrank}(\widetilde{\Omega\mathcal{P}^+}(M))\\ &:= \sup\{\mathrm{rank}(\mathrm{im}(\mathrm{R}_{n+1}^{\mathrm{spin}}(M)_U \to \mathrm{Pos}_n^{\mathrm{spin}}(M)_U)) \mid U \le \mathrm{Diffeo}(M) \text{ of finite index}\}.\end{aligned}$$

REMARK 11.25. Obviously, whenever $\operatorname{vrank}(\widetilde{\mathcal{P}^+}(M)) > 0$, then

$$\left|\pi_0(\operatorname{Riem}^+(M)/\operatorname{Diffeo}(M))\right| = \infty.$$

This follows because there is a finite index subgroup $U \subset \operatorname{Diffeo}(M)$ such that $\pi_0(\operatorname{Riem}^+(M))/U$ surjects onto $\mathcal{P}^+(M)/U$ which maps surjectively to an abelian group $\mathrm{R}^{\mathrm{spin}}_{n+1}(M)_U$ of positive rank. In particular, all the sets involved are infinite.

As U has finite index in $\operatorname{Diffeo}(M)$, then also the full quotient

$$\pi_0(\operatorname{Riem}^+(M)/\operatorname{Diffeo}(M)) = \pi_0(\operatorname{Riem}^+(M))/\operatorname{Diffeo}(M)$$

has to be infinite.

REMARK 11.26. After choices of basepoints, the above maps define non-natural abelian group structures on quotients $\mathcal{P}^+(M)/U$ (by identifying them with $\mathrm{R}^{\mathrm{spin}}_{n+1}(M)_U$ or with a subgroup of $\mathrm{Pos}^{\mathrm{spin}}_n(M)_U$). In [**81**, Definition 5.7] the set $\widetilde{\mathcal{P}^+(M)}$ is defined for the first time (using the full diffeomorphisms group $\operatorname{Diffeo}(M)$), together with a non-canonical abelian group structure depending on the choice of a basepoint. This group structure coincide with the one we are using, as shown in [**84**, Proposition 4.1].

Although not stated there explicitly, also [**81**] employ the concept of vrank: there one has to switch from the full diffeomorphism group $\operatorname{Diffeo}(M)$ to the spin structure preserving subgroup U to obtain compatibility of the rho invariant with the action and then get a well defined map $\tilde{\varrho}\colon \widetilde{\mathcal{P}^+}(M) \to \mathrm{S}^\Gamma_n(\widetilde{M})_U$.

We will apply our techniques to obtain lower bounds on the virtual concordance moduli space and bordism moduli space of Riemannian metrics of positive scalar curvature on suitable manifolds M.

CHAPTER 12

The role of real K-theory

In this paper, for simplicity we have worked throughout with complex K-theory. To get more precise information on the scope of the applicability to the classification of positive scalar curvature metrics, we must observe that the Stolz positive scalar curvature exact sequence actually maps to a real version of the Higson-Roe analytic surgery exact sequence [**87**, Theorem 3.1.13], which we write as

$$\cdots \to KO_{*+1}(C^*_{\mathbb{R},red}\Gamma) \xrightarrow{\iota} \mathrm{SO}^\Gamma_*(\widetilde{M}) \to KO_*(M) \xrightarrow{\mu_{\mathbb{R}}} KO_*(C^*_{\mathbb{R},red}\Gamma) \to \dots \tag{12.1}$$

To understand the structure group $\mathrm{SO}^\Gamma_*(\widetilde{M})$ which is our main object of study, in light of the exact sequence it is important to understand the KO-theory of the group C^*-algebra.

The Baum-Connes conjecture asserts that the assembly map μ is an isomorphism, known for large classes of discrete groups, e.g. for hyperbolic groups:

$$KO^\Gamma_*(\underline{E}\Gamma) \xrightarrow{\mu} KO_*(C^*_{\mathbb{R},red}\Gamma). \tag{12.2}$$

Real and complex K-theory are connected by a long exact sequence which canonically splits after inverting 2 (compare e.g. [**69**, Theorem 2.1]), giving in particular the split exact sequence

$$0 \to KO_p(C^*_{\mathbb{R},red}\Gamma) \otimes \mathbb{Q} \xrightarrow{c} K_p(C^*_{red}\Gamma) \otimes \mathbb{Q} \xrightarrow{\delta} KO_{p-2}(C^*_{\mathbb{R},red}\Gamma) \otimes \mathbb{Q} \to 0. \tag{12.3}$$

Here, c is complexification and δ is the composition of the Bott periodicity isomorphism with "forgetting the complex structure". In other words, $K_k(C^*\Gamma) \otimes \mathbb{Q}$ is canonically the direct sum of real K-theory in degree k and in degree $k-2$.

After complexification, we have a Chern character isomorphism

$$K^\Gamma_p(\underline{E}\Gamma) \otimes \mathbb{C} \xrightarrow[\cong]{\mathrm{Ch}} \bigoplus_{k \in \mathbb{Z}} H_{p+2k}(\Gamma; F\Gamma), \tag{12.4}$$

worked out in great detail in [**57**, Theorem 1.4] or with a different model of K-homology in [**15**, Section 4, 5]. Here, $F\Gamma := \{f \colon \Gamma_{fin} \to \mathbb{C} \mid |\mathrm{supp}(f)| < \infty\}$, where Γ_{fin} is the set of elements of finite order in Γ. It becomes a Γ-module by conjugation. Note that $F\Gamma$ is a direct sum of permutation modules $F_{[\gamma]}\Gamma := \{f \in F\Gamma \mid \mathrm{supp}(f) \subset \langle\gamma\rangle\}$, one for each conjugacy class $[\gamma]$ of elements of finite order $\gamma \in \Gamma$. The centralizer Γ_γ is the stabilizer group of γ under the conjugation action, and the Γ-module $F_{[\gamma]}\Gamma$ is precisely the trivial Γ_γ-module $\mathbb{C}$ induced up to Γ, $F_{[\gamma]}\Gamma = \mathrm{ind}_{\Gamma_\gamma}^\Gamma \mathbb{C}$. It follows that we get

$$F\Gamma = \bigoplus_{[\gamma], \gamma \in \Gamma \text{ of finite order}} F_{[\gamma]}\Gamma = \bigoplus_{[\gamma], \gamma \in \Gamma \text{ of finite order}} \mathrm{ind}_{\Gamma_\gamma}^\Gamma \mathbb{C}. \tag{12.5}$$

Recall that group homology is compatible with induction: for a subgroup $H \subset \Gamma$ we have a canonical isomorphism $H_*(\Gamma; \mathrm{ind}_H^\Gamma F) \xrightarrow{\cong} H_*(H; F)$. In the present case we therefore get, combining this and the decomposition (12.5)

$$\begin{aligned} H_*(\Gamma; F_{[\gamma]\Gamma}) &= H_*(\Gamma, \operatorname{ind}_{\Gamma_\gamma}^{\Gamma} \mathbb{C}) \xrightarrow{\cong} H_*(\Gamma_\gamma; \mathbb{C}), \\ H_*(\Gamma; F\Gamma) &= \bigoplus_{[\gamma], \gamma\in\Gamma \text{ of finite order}} H_*(\Gamma_\gamma; \mathbb{C}). \end{aligned} \tag{12.6}$$

See also [**7**, (15.11)]. Note that these are precisely the summands in the computation of periodic cyclic homology as derived by Burghelea [**13**], as explained in the proof of Corollary 7.9. Even more is true, there is a canonical inclusion $H_*(\Gamma; F\Gamma) \hookrightarrow HP_*(\mathbb{C}\Gamma)$ which is an isomorphism for all groups where conjugacy classes of elements of infinite order don't contribute to $HP_*(\mathbb{C}\Gamma)$, e.g. for hyperbolic groups or groups of polynomial growth.

With this canonical map, we get a well known commutative diagram connecting the Chern character on the topological side of the Baum-Connes assembly map to the Chern character on the operator algebra side

$$\begin{array}{ccccc} K_p^{\Gamma}(\underline{E}\Gamma) & \longrightarrow & & & K_p(C^*_{red}\Gamma) \\ \downarrow \text{Ch} & & & & \downarrow \text{Ch} \\ \bigoplus_{k\in\mathbb{Z}} H_{p+2k}(\Gamma; F\Gamma) & \longrightarrow & HP_p(\mathbb{C}\Gamma) & \longrightarrow & HP_p(\mathcal{A}\Gamma) \end{array} \tag{12.7}$$

where as usual $\mathcal{A}\Gamma$ is a dense and holomorphically closed subalgebra of Γ, and $HP_*(\mathcal{A}\Gamma)$ shows up as explained in Section 7.1.

Let us finally define $F_{del}\Gamma := \{f \in F\Gamma \mid \sum f(g) = 0\}$. This is a subrepresentation of ΓF which is complementary to the sub-representation of functions supported on e. The latter is of course the trivial representation and the permutation module $F_{[e]}\Gamma$. $F_{del}\Gamma$ therefore is the sum of all permutation modules for the conjugacy classes of non-trivial elements.

Under the monomorphism $H_*(\Gamma; F\Gamma) \to HP_*(\mathbb{C}\Gamma)$, the image of $H_*(\Gamma; F_{[e]}\Gamma)$ is precisely the localized periodic cyclic homology $HP_*^e(\mathbb{C}\Gamma)$ and $H_*(\Gamma; F_{del}\Gamma)$ maps to delocalized periodic cyclic homology $HP_*^{del}(\mathbb{C}\Gamma)$, even isomorphically, if no conjugacy class of an element of infinite order contributes non-trivially to $HP_*(\mathbb{C}\Gamma)$.

The splitting of complex K-theory into a sum of two (shifted) copies of real K-theory of (12.3) also causes the Chern character for complex K-theory (12.4) to split into a Chern character for the real K-theory summands, sometimes called Pontryagin character. This is worked out in [**6**, Proposition 2.2] and gives

$$KO_p^{\Gamma}(\underline{E}\Gamma) \otimes \mathbb{C} \xrightarrow[\cong]{\text{Ch}} \bigoplus_{k\in\mathbb{Z}} H_{p+4k}(\Gamma; F^0\Gamma) \oplus H_{p+2+4k}(\Gamma; F^1\Gamma). \tag{12.8}$$

Here, we need a somewhat more elaborate splitting of the homology. Observe that $F\Gamma$ is decomposed as the direct sum $F\Gamma = F^0\Gamma \oplus F^1\Gamma$ with $F^r\Gamma = \{f \in F\Gamma \mid f(\gamma) = (-1)^r f(\gamma^{-1})\}$, the ± 1-eigenspaces for the involution $\tau\colon F\Gamma \to F\Gamma$ induced by sending γ to γ^{-1}. We want to combine this with the decomposition of $F\Gamma$ into permutation modules $F_{[\gamma]}\Gamma$. Unfortunately, this is a bit more involved as not all the $F_{[\gamma]}\Gamma$ are preserved by τ. To sort this out we make the following observations and definitions.

LEMMA 12.1. *For $\gamma \in \Gamma$ of finite order, let $F_{\{[\gamma],[\gamma^{-1}]\}}\Gamma$ be the smallest $\mathbb{C}\Gamma$ submodule containing $F_{[\gamma]}\Gamma$ closed under τ. We then get the following structure result as $\mathbb{C}\Gamma$ modules.*

- *If $\gamma\in\Gamma$ is of finite order and not conjugate to its inverse, then $F_{\{[\gamma],[\gamma^{-1}]\}} = F_{[\gamma]}\Gamma \oplus F_{[\gamma^{-1}]}\Gamma$ split under the involution τ into an even and an odd summand, both isomorphic to $F_{[\gamma]}\Gamma$ as Γ-modules.*
- *If $\gamma = \gamma^{-1}$, i.e. $\gamma^2 = 1$, then τ acts trivially on $F_{\{[\gamma],[\gamma^{-1}]\}}\Gamma = F_{[\gamma]}\Gamma$ and this module belongs entirely to $F^0\Gamma$.*
- *If $\gamma \neq \gamma^{-1}$ but the two elements are conjugate, let $\Gamma_{\{\gamma,\gamma^{-1}\}}$ be the subgroup of Γ conjugating the set $\{\gamma,\gamma^{-1}\}$ to itself. It contains Γ_γ as normal subgroup of index 2. The Γ-module $F_{\{[\gamma],[\gamma^{-1}]\}}\Gamma = F_{[\gamma]}\Gamma$ can be considered as induced up from the trivial Γ_γ-module or, equivalently, from the $\Gamma_{\{\gamma,\gamma^{-1}\}}$-module $\mathbb{C}[\Gamma_{\{\gamma,\gamma^{-1}\}}/\Gamma_\gamma]$. The latter decomposes into τ-eigenspaces as the trivial representation of $\Gamma_{\{\gamma,\gamma^{-1}\}}$ for the even part and the sign representation $\mathbb{C}^{\mathrm{sgn}}$ (via the projection to the 2-element group $\Gamma_{\{\gamma,\gamma^{-1}\}}/\Gamma_\gamma$) as the odd part. Therefore, it contributes a summand of the form $H_*(\Gamma_{\{\gamma,\gamma^{-1}\}};\mathbb{C})$ to $H_*(\Gamma;F^0\Gamma)$ and a summand of the form $H_*(\Gamma_{\{\gamma,\gamma^{-1}\}};\mathbb{C}^{\mathrm{sgn}})$ to $H_*(\Gamma;F^1\Gamma)$.*

PROOF. These results are certainly well known and easy exercises. For the convenience of the reader and in lack of an appropriate reference, we give the arguments.

If γ is not conjugate to its inverse, then $F_{[\gamma]}\Gamma \cap F_{[\gamma^{-1}]} = \{0\}$ and we therefore have the direct sum decomposition $F_{\{[\gamma],[\gamma^{=1}]\}} = F_{[\gamma]}\Gamma \oplus F_{[\gamma^{-1}]}\Gamma$ and τ just exchanges the two summands. We have then $\mathbb{C}\Gamma$-module isomorphisms from $F_{[\gamma]}\Gamma$ to the ±1-eigenspaces of the action of τ on $F_{[\gamma],[\gamma^{-1}]}\Gamma$ given explicitly by

$$F_{[\gamma]}\Gamma \xrightarrow{\cong} F^0_{\{[\gamma],[\gamma^{=1}]\}}\Gamma; x \mapsto \frac{1}{2}(x,x)$$
$$F_{[\gamma]}\Gamma \xrightarrow{\cong} F^1_{\{[\gamma],[\gamma^{=1}]\}}\Gamma; x \mapsto \frac{1}{2}(x,-x)$$

If $\gamma = \gamma^{-1}$ then $\tau(g\gamma g^{-1}) = (g\gamma g^{-1})^{-1} = g\gamma g^{-1}$ i.e. τ acts trivially on $F_{[\gamma]}$, which is therefore also equal to $F_{\{[\gamma],[\gamma^{-1}]\}}$ and belongs entirely to the $+1$-eigenspace $F^0\Gamma$.

Finally, if $\gamma \neq \gamma^{-1}$ but the two elements are conjugate, then τ preserves $F_{[\gamma]\Gamma}$ and therefore $F_{\{[\gamma],[\gamma^{-1}]\}}\Gamma = F_{[\gamma]}\Gamma$. We need to identify the ±1-eigenspaces for the action of τ on $F_{[\gamma]}\Gamma$ as $\mathbb{C}\Gamma$-modules.

Note that the subgroup $\Gamma_{\{\gamma,\gamma^{-1}\}}$ acts transitively on $\{\gamma,\gamma^{-1}\}$ because we assume that γ and γ^{-1} are conjugate and therefore surjects onto the permutation group of this set, which is the 2-element group. The kernel of this surjection consists of those group elements which conjugate γ to itself, and then automatically also γ^{-1} to γ^{-1}, i.e. of the centralizer Γ_γ. We get the desired extension

$$1 \to \Gamma_\gamma \to \Gamma_{\{\gamma,\gamma^{-1}\}} \to \mathbb{Z}/2\mathbb{Z} \to 1. \tag{12.9}$$

Because $F_{[\gamma]}\Gamma$ is a permutation module via the transitive action of Γ on $[\gamma]$ (by conjugation) and with stabilizer group Γ_γ of the element $\gamma \in [\gamma]$, we have the standard identification as induced representation (the $=$ is an isomorphism which depends on the choice of the basepoint $\gamma \in [\gamma]$) $F_{[\gamma]}\Gamma = \mathrm{ind}_{\Gamma_\gamma}^{\Gamma}\mathbb{C}$. By transitivity of induction of representation, we can write this as

$$F_{[\gamma]}\Gamma = \mathrm{ind}_{\Gamma_{\{\gamma,\gamma^{-1}\}}}^{\Gamma}(\mathrm{ind}_{\Gamma_\gamma}^{\Gamma_{\{\gamma,\gamma^{-1}\}}}\mathbb{C}).$$

By the explicit description of induced representations, $\mathrm{ind}_{\Gamma_\gamma}^{\Gamma_{\{\gamma,\gamma^{-1}\}}} \mathbb{C}$ is the $\Gamma_{\{\gamma,\gamma^{-1}\}}$-permutation module $\mathbb{C}[\Gamma_{\{\gamma,\gamma^{-1}\}}/\Gamma_\gamma]$ where the group $\Gamma_{\{\gamma,\gamma^{-1}\}}$ acts by left multiplication on the coset "space" $\Gamma_{\{\gamma,\gamma^{-1}\}}/\Gamma_\gamma$. Note that this action is the permutation action on the 2-element set via the projection $\Gamma_{\{\gamma,\gamma^{-1}\}} \to \mathbb{Z}/2\mathbb{Z}$ of the extension (12.9).

To determine the action of τ in this situation, choose $g \in \Gamma_{\{\gamma,\gamma^{-1}\}} \setminus \Gamma_\gamma$, so that we have the right coset decomposition $\Gamma_{\{\gamma,\gamma^{-1}\}} = \Gamma_\gamma \sqcup g\Gamma_\gamma$. Note that g acts non-trivially on $\{\gamma, \gamma^{-1}\}$ by conjugation, so $\gamma^{-1} = g\gamma g^{-1}$. Moreover, the ± 1-eigenspaces of $F_{[\gamma]}\Gamma$ are generated as $\mathbb{C}\Gamma$-modules by $\gamma \pm \gamma^{-1} = \gamma \pm g\gamma g^{-1}$. Comparing with the definition of the induced representation, this implies that the sub-representation $\Gamma^0_{[\gamma]}$ is induced up from the trivial sub-representation $\mathbb{C}$ of $\Gamma_{\{\gamma,\gamma^{-1}\}}$, while $\Gamma^1_{[\gamma]}$ is induced up from the sign representation $\mathbb{C}^{\mathrm{sgn}}$, where the sign representation is defined via the projection $\Gamma_{\{\gamma,\gamma^{-1}\}} \to \mathbb{Z}/2$ of (12.9).

Finally, the standard isomorphism of group homology with coefficients in induced representations gives in the present case

$$\begin{aligned} H_*(\Gamma; F^0_{[\gamma]}\Gamma) &= H_*(\Gamma; \mathrm{ind}_{\Gamma_{\{\gamma,\gamma^{-1}\}}}^{\Gamma} \mathbb{C}) = H_*(\Gamma_{\{\gamma,\gamma^{-1}\}}; \mathbb{C}),\\ H_*(\Gamma; F^1_{[\gamma]}\Gamma) &= H_*(\Gamma; \mathrm{ind}_{\Gamma_{\{\gamma,\gamma^{-1}\}}}^{\Gamma} \mathbb{C}^{\mathrm{sgn}}) = H_*(\Gamma_{\{\gamma,\gamma^{-1}\}}; \mathbb{C}^{\mathrm{sgn}}). \end{aligned} \qquad \square$$

Observe that in the cases γ not conjugate to γ^{-1} or $\gamma = \gamma^{-1}$ of Lemma 12.1 the resulting modules $F^0_{[\gamma]}\Gamma$ and $F^1_{[\gamma]}\Gamma$ are obviously induced up from the trivial (or zero) Γ_γ-module. We therefore have a complete and explicit decomposition of the target of the Pontryagin character for real K-theory, refining (12.6).

Corollary 12.2. *The Pontryagin character* (12.8) *can be written as an isomorphism*

(12.10)

$$KO_p^\Gamma(\underline{E}\Gamma) \otimes \mathbb{C} \xrightarrow[\cong]{\mathrm{Ch}}$$

$$\bigoplus_{k\in\mathbb{Z}} \left(\bigoplus_{[\gamma]\neq[\gamma^{-1}] \text{ or } \gamma=\gamma^{-1}} H_{p+4k}(\Gamma_\gamma; \mathbb{C}) \oplus \bigoplus_{[\gamma]=[\gamma^{-1}] \text{ and } \gamma\neq\gamma^{-1}} H_{p+4k}(\Gamma_{\{\gamma,\gamma^{-1}\}}; \mathbb{C}) \right)$$

$$\oplus \bigoplus_{k\in\mathbb{Z}} \left(\bigoplus_{[\gamma]\neq[\gamma^{-1}]} H_{p+2+4k}(\Gamma_\gamma; \mathbb{C}) \oplus \bigoplus_{[\gamma]=[\gamma^{-1}] \text{ and } \gamma\neq\gamma^{-1}} H_{p+2+4k}(\Gamma_{\{\gamma,\gamma^{-1}\}}; \mathbb{C}^{\mathrm{sgn}}) \right)$$

Here, the sums are over all conjugacy classes $[\gamma]$ of elements $\gamma \in \Gamma$ of finite order (satisfying the additional conditions as stated in the formula).

The dual cohomology groups decompose accordingly (as direct products instead of direct sums), and the homology/cohomology pairing respects this decomposition, i.e. each homology summand pairs non-trivially only with the corresponding cohomology summand and these individual pairings are non-degenerate. For later reference, let us also spell this out explicitly:

(12.11)

$$\bigoplus_{k\in\mathbb{Z}} H^{p+4k}(\Gamma;F^0\Gamma)\oplus H^{p+2+4k}(\Gamma;F^1\Gamma)$$

$$\cong \bigoplus_{k\in\mathbb{Z}}\left(\prod_{[\gamma]\neq[\gamma^{-1}]\ or\ \gamma=\gamma^{-1}} H^{p+4k}(\Gamma_\gamma;\mathbb{C})\oplus \prod_{[\gamma]=[\gamma^{-1}]\ and\ \gamma\neq\gamma^{-1}} H^{p+4k}(\Gamma_{\{\gamma,\gamma^{-1}\}};\mathbb{C})\right)$$

$$\oplus \bigoplus_{k\in\mathbb{Z}}\left(\prod_{[\gamma]\neq[\gamma^{-1}]} H^{p+2+4k}(\Gamma_\gamma;\mathbb{C})\oplus \prod_{[\gamma]=[\gamma^{-1}]\ and\ \gamma\neq\gamma^{-1}} H^{p+2+4k}(\Gamma_{\{\gamma,\gamma^{-1}\}};\mathbb{C}^{\mathrm{sgn}})\right)$$

Under the decomposition of complex K-theory as direct sum of two shifted copies of real K-theory $K_*\otimes\mathbb{C}\cong KO_*\otimes\mathbb{C}\oplus KO_{*-2}\otimes\mathbb{C}$, the complex Baum-Connes assembly map splits as the direct sum of two shifted copies of the real Baum-Connes assembly map (12.2), and also the group cohomology $H^*(\Gamma;F\Gamma)$ which pairs with $K_*^\Gamma(\underline{E}\Gamma)$ splits compatibly, as described in (12.10).

Note that this description is compatible with the computation of periodic cyclic group cohomology in Corollary 7.9, going back to [**13**]. Specifically, we have a real version of the diagram (12.7) as follows

(12.12)

$$\begin{array}{ccc} KO_p^\Gamma(\underline{E}\Gamma) & \longrightarrow & KO_p(C^*_{\mathbb{R},red}\Gamma) \\ \downarrow{\scriptstyle \mathrm{Ch}} & & \downarrow{\scriptstyle \mathrm{Ch}} \\ \bigoplus_{k\in\mathbb{Z}} H_{p+4k}(\Gamma;F^0\Gamma)\oplus H_{p+2+4k}(\Gamma;F^1\Gamma) & \longrightarrow & HP_p(\mathcal{A}\Gamma)^0\oplus HP_p(\mathcal{A}\Gamma)^1 \end{array}$$

where the superscripts 0 and 1 of the periodic cyclic homology groups denote the ±1-eigenspaces under the endomorphism of cyclic homology induced by mapping $\gamma\in\Gamma$ to γ^{-1}.

The pairings with the summands $H^*(\Gamma;F^{0/1}_{\{[\gamma],[\gamma^{-1}]\}}\Gamma)$ of $HP^*(\mathbb{C}\Gamma)$ on the right hand side are compatible with the pairings of the same summands of $H^*(\Gamma;F\Gamma)$ with the left hand side.

Summarizing, if Γ is hyperbolic or of polynomial growth, the dual of the bottom line in (12.12) tells us that we have an isomorphism

$$HP^*_{del}(\mathbb{C}\Gamma)\to H^*(\Gamma,F_{del}\Gamma) \tag{12.13}$$

and that, for $\gamma\in\Gamma$ of finite order, $H^*(\Gamma_\gamma;\mathbb{C})$ is isomorphic to a summand, in a compatible way, of both members of (12.13). We obtain in this way the following commutative diagram

(12.14)

$$\begin{array}{ccccc} & & H^{*+4k+2\epsilon}(\Gamma_\gamma;\mathbb{C}) & & \\ & \swarrow & \downarrow & \searrow & \\ HP^*_{del}(\mathbb{C}\Gamma) & \xrightarrow{=} & HP^*_{del}(\mathbb{C}\Gamma) & \longrightarrow & H^*(\Gamma,F_{del}\Gamma) \\ \downarrow & & \downarrow & & \downarrow \\ \mathrm{SO}^\Gamma_{*-1}(\widetilde{M})' & \longrightarrow & KO_*(C^*_{\mathbb{R},red}\Gamma)' & \longrightarrow & KO^\Gamma_*(\underline{E}\Gamma)' \end{array}$$

for $\gamma \in \Gamma$ of finite order and $[\gamma] \neq [\gamma^{-1}]$, $k \in \mathbb{Z}$ and $\epsilon \in \{0, 1\}$; a similar diagram can be written for the other $\gamma \neq e$ summands of (12.10).

CHAPTER 13

Geometric applications of delocalized cyclic cohomology

We are now in the situation to formulate a meta theorem which describes one of the main geometric application of our higher rho numbers, which help to distinguish metrics of positive scalar curvature.

First, observe that for $\psi \in \mathrm{Aut}(\Gamma)$ the Γ-module $\psi^* F\Gamma$ can be canonically identified with $F\Gamma$, and therefore indeed $\mathrm{Out}(\Gamma)$ acts on $H^*(\Gamma; F\Gamma)$, the action also preserves the decomposition in the even and odd part relevant in the Chern character for real K-theory.

THEOREM 13.1. *Let M be a closed n-dimensional spin manifold with a Riemannian metric g of positive scalar curvature. Set $\Gamma = \pi_1(M)$. We have the KO-theoretic rho-invariant $\varrho(g) \in \mathrm{SO}_*^\Gamma(\widetilde{M})$. Assume there is a subspace $H \subset H^*(\Gamma; F_{del}\Gamma)$ of dimension $k > 0$ which is pointwise fixed by a finite index subgroup V of* $\mathrm{Out}(\Gamma)$.

Assume that there is a dense and holomorphically closed Fréchet subalgebra $\mathcal{A}\Gamma$ of $C^*_{red}\Gamma$ which contains $\mathbb{C}\Gamma$ and to which all $\lambda \in H$ extend. By the construction of Sections 6 *and* 7.6*, specifically by Definition* 7.24*, the subgroup H then pairs with* $\mathrm{SO}_n^\Gamma(\widetilde{M})$.

Assume that we can find a subgroup $B \subset KO_{n+1}(C^*_{\mathbb{R},red}\Gamma)$ such that the pairing between B and H is non-degenerate (in particular, $\mathrm{rank}(B) \geq k$*) and assume that we can construct metrics g_β on M of positive scalar curvature for all $\beta \in B$ such that*

$$\varrho(g_\beta) = \varrho(g) + \iota(\beta) \tag{13.1}$$

with ι as in (12.1)*. Then the metrics $\{g_\beta \mid \beta \in B\}$ represent infinitely many different components of the moduli space of metrics of positive scalar curvature of M. More precisely, their images in* $\mathrm{Pos}_n^{\mathrm{spin}}(M)$ *span a coset of rank k whose image in the coinvariant group* $\mathrm{Pos}_n^{\mathrm{spin}}(M)_U$ *for a suitable finite index subgroup U of* $\mathrm{Diffeo}(M)$ *still has rank k.*

In particular, $\mathrm{vrank}(\widetilde{\mathcal{P}^+}(M)) \geq k$ *and even* $\mathrm{vrank}(\widetilde{\Omega\mathcal{P}^+}(M)) \geq k$*, the virtual rank of the concordance moduli space and even of the bordism moduli space of metrics of positive scalar curvature is $\geq k$.*

PROOF. By naturality of the pairings (12.14) and linearity, we have

$$\langle \varrho(g_\beta), \lambda \rangle - \langle \varrho(g_{\beta'}), \lambda \rangle = \langle \beta - \beta', \lambda \rangle \qquad \forall \beta \in B, \lambda \in H. \tag{13.2}$$

Recall that the subgroup $\mathrm{Diffeo}^{\mathrm{spin}}(M)$ of spin structure preserving diffeomorphisms of M has finite index in $\mathrm{Diffeo}(M)$. Let U be the intersection of $\mathrm{Diffeo}^{\mathrm{spin}}(M)$ and the inverse image of V under $\alpha\colon \mathrm{Diffeo}(M) \to \mathrm{Out}(\Gamma)$, which is then also a finite index subgroup of $\mathrm{Diffeo}(M)$.

Because H is invariant under U and U consists of spin structure preserving diffeomorphisms, its pairing with $\mathrm{Pos}_n^{\mathrm{spin}}(M)$ descends to a pairing with the coinvariants

$$H \times \mathrm{Pos}_n^{\mathrm{spin}}(M)_U \to \mathbb{C}.$$

Indeed, for a diffeomorphism ψ and a metric g of positive scalar curvature on M we have

$$(13.3) \qquad \langle \varrho(\psi^* g_\beta), \lambda\rangle = \langle \varrho(g_\beta), (\alpha_\psi^{-1})^*\lambda\rangle = \langle \varrho(g_\beta), \lambda\rangle \quad \forall \beta \in B, \lambda \in H.$$

Here we use (11.6) for the first equality and for the second one we use our invariance assumption on H.

By (13.2), on the affine subgroup spanned by the g_β for $\beta \in B$, the pairing with H is at least as non-degenerate as the pairing of B with H and therefore the rank of this affine subgroup is at least k.

Finally, because $U \subset \mathrm{Diffeo}(M)$ has finite index, $\mathrm{Riem}^+(M)/\mathrm{Diffeo}(M)$ is a further finite quotient of the infinite set $\mathrm{Riem}^+(M)/U$ and therefore the classes of the g_β still lie in infinitely many different components. □

We now give specific examples for the meta theorem. We start with auxiliary results which guarantee the existence of cocycles with the desired invariance property.

The most obvious one is that the outer automorphism group is finite. In this case, H in Theorem 13.1 can be taken to be equal to $H^*(\Gamma; F_{del}\Gamma)$. Generically, this is the case for hyperbolic groups, as proven by Bestvina and Feighn, refining a result of Paulin:

PROPOSITION 13.2 ([**10**, Corollary 1.3]). *Let Γ be a hyperbolic group. Then $|\mathrm{Out}(\Gamma)| = \infty$ if and only if Γ does split over a virtually cyclic group*

Recall here that a group Γ is defined to *split over a subgroup* Λ if Γ is a non-trivial amalgamated product, amalgamated over Λ, or an HNN extension with base group Λ.

EXAMPLE 13.3. Clearly, non-abelian free groups split over the trivial group, and hyperbolic surface groups split over infinite cyclic groups, and their outer automorphism groups are infinite.

On the other hand, hyperbolic fundamental groups of aspherical manifolds don't split over virtually cyclic subgroups (and have finite outer automorphism group), compare [**27**, §5, 5.4.A]. These groups are torsion free, but then also any finite extension Γ of such a group doesn't split and therefore $|\mathrm{Out}(\Gamma)| < \infty$.

Our second class of examples makes use of the specific structure of the delocalized cyclic group cohomology to identify subspaces which necessarily are fixed under a finite index subgroup of the outer automorphism group, without making any a priori assumption on the latter. For this, it is important to observe that the action of $\mathrm{Diffeo}(M)$ on $HP^*(\mathbb{C}\Gamma)$ (which factors through the action of $\mathrm{Out}(\Gamma)$) can be refined. Recall the splitting of $HP^*(\mathbb{C}\Gamma)$ given in (12.11). The naturality of the isomorphism implies that the action of $\mathrm{Out}(\Gamma)$ on $HP^*(\mathbb{C}\Gamma)$ becomes the evident action on this direct product: $\alpha \in \mathrm{Aut}(\Gamma)$ maps Γ_γ to $\Gamma_{\alpha(\gamma)}$ and this map induces a well defined action of $\mathrm{Out}(\Gamma)$ on the cohomology groups forming the summands of the decomposition (12.11). In other words, $\alpha_\psi \in \mathrm{Out}(\Gamma)$ acts by permuting the summands corresponding to different conjugacy classes the same way α_ψ permutes the conjugacy classes.

LEMMA 13.4. *Assume that, in the decomposition* (12.11) *dual to the right hand side of the Pontryagin character, one of the summands* H, *e.g. the group cohomology of one of the centralizers in one fixed degree, satisfies* $\dim(H) = 1$.

Assume that $|\mathrm{Out}(\Gamma) \cdot [\gamma]| < \infty$, *i.e. applying all automorphisms of* Γ *to* γ *yields only finitely many new conjugacy classes. Then a subgroup of finite index of* $\mathrm{Out}(\Gamma)$ *acts trivially on the summand* H.

PROOF. By assumption, the $\mathrm{Out}(\Gamma)$-orbit of the conjugacy class $[\gamma]$ is finite, therefore its stabilizer $\mathrm{Out}(\Gamma)_{[\gamma]} \subset \mathrm{Out}(\Gamma)$ has finite index in $\mathrm{Out}(\Gamma)$.

Finally, the summand H of (12.10), which is a fixed cohomology group of a particular subgroup of Γ, contains the image of the cohomology with $\mathbb{Z}$-coefficients in cohomology with $\mathbb{C}$-coefficients as an integer lattice. This lattice is isomorphic to $\mathbb{Z}$, because $\dim(H) = 1$ by assumption. Moreover, the action of $\mathrm{Out}(\Gamma)_{[\gamma]}$ on H factors through the automorphisms of this integral lattice, i.e. through $\mathbb{Z}/2 = \mathrm{Aut}(\mathbb{Z})$. It follows that a subgroup of index 1 or 2 acts trivially on H. Therefore also a subgroup of index 1 or 2 of $\mathrm{Diffeo}(M)_{[\gamma]}$ acts trivially on H. Its intersection with the spin structure preserving diffeomorphisms is the desired subgroup U of finite index. □

LEMMA 13.5. *If* Γ *is hyperbolic or of polynomial growth then* Γ *has only finitely many conjugacy classes of elements of finite order, in particular, for each* $\gamma \in \Gamma$ *of finite order,* $|\mathrm{Out}(\Gamma) \cdot [\gamma]| < \infty$.

PROOF. It is well known that hyperbolic groups have only finitely many conjugacy classes of elements of finite order, compare e.g. [**58**, Corollary 8].

Groups of polynomial growth contain a subgroup of finite index which is finitely generated nilpotent and therefore such groups are in particular virtually poly-cyclic. The latter groups are known to have a finite model for the universal space for proper actions and in particular have only finitely many conjugacy classes of finite order [**55**]. □

COROLLARY 13.6. *Assume that* Γ *is hyperbolic. Then the direct sum of all summands in the decomposition dual to* (12.10) *which are* 1*-dimensional forms a subspace* $H \subset H^*(\Gamma; F_{del}\Gamma)$ *which is pointwise fixed by a finite index subgroup of* $\mathrm{Out}(\Gamma)$.

EXAMPLE 13.7. Let Γ be a hyperbolic group or a group of polynomial growth. Then the direct sum of all summands of homological degree 0 in the decomposition of cyclic group cohomology (12.10) forms a subspace H which is pointwise fixed by a finite index subgroup of $\mathrm{Out}(\Gamma)$. Let p be as in Corollary 12.2. For $p \equiv 0 \pmod 4$ its rank is equal to the number of conjugacy classes of elements of finite order. For $p = 2$ its rank is equal to the number of conjugacy classes of elements which are not conjugate to their inverse.

EXAMPLE 13.8. Assume that Γ_0 is a hyperbolic group with unit e_0. Assume for $0 = k_1 < k_2 < \cdots < k_r$ that $\dim H^{k_i}(\Gamma_0; \mathbb{C}) = 1$. As a specific example, let N be a closed connected orientable manifold of dimension k_2 with negative sectional curvature and set $\Gamma_0 := \pi_1(N)$. Let F be a non-trivial finite group and set $\Gamma := \Gamma_0 \times F$.

Then, for each $e \neq \gamma \in F$, its centralizer $\Gamma_{(e_0,\gamma)}$ surjects onto Γ_0 with a finite kernel (a subgroup of F). In particular, $H^*(\Gamma_{(e_0,\gamma)}; \mathbb{C}) \cong H^*(\Gamma_0; \mathbb{C})$, and therefore $\dim H^{k_i}(\Gamma_{(e_0,\gamma)}; \mathbb{C}) = 1$ by the assumption on Γ.

Therefore, the $H^{k_i}(\Gamma_{(e_0,\gamma)};\mathbb{C})$ provide summands H of the decomposition (12.10) as in Lemma 13.4, and we obtain a subspace of delocalized cyclic group cohomology $H \subset H^*(\Gamma; F_{del}\Gamma)$ of dimension rd which is fixed by a finite index subgroup of $\mathrm{Out}(\Gamma)$, with d the number of non-trivial conjugacy classes in F.

These results provide an ample supply of examples as required in the meta theorem 13.1. It remains to discuss the existence of the metrics g_k which can be distinguished from each other as in Equation (13.1). This is achieved in [**84**, Corollary 1.2], or more precisely in the proof of [**84**, Theorem 1.1], compare also [**72**, Lemma 2.1]. From these references, we obtain the following theorem.

THEOREM 13.9. *Let M be a closed connected Riemannian spin manifold of dimension $n \geq 5$ with a Riemannian metric g of positive scalar curvature. Assume that $\Gamma = \pi_1(M)$ satisfies the rational strong Novikov conjecture, i.e. the Baum-Connes assembly map $\mu_{\mathbb{R}}\colon KO_*^{\Gamma}(\underline{E}\Gamma) \to KO_*(C^*_{\mathbb{R},red}\Gamma)$ is rationally injective.*

Then for each subgroup $H \subset \bigoplus_{k>0,p\in\{0,1\}} H^{n+1-4k-2p}(\Gamma; F^p_{del}\Gamma)$ (which can be translated to group cohomology of centralizers of elements of finite order as in (12.10)) such that the classes in H are represented by cocycles which extend to a smooth subalgebra $\mathcal{A}\Gamma$ of $C^*_{red}\Gamma$, there is a subgroup $B \subset KO_{n+1}(C^*_{\mathbb{R},red}\Gamma)$ and for all $\beta \in B$ there are metrics g_β of positive scalar curvature on M such that

$$\varrho(g_\beta) = \varrho(g) + \iota(\beta); \qquad \forall \beta \in B$$

and such that the pairing between H and B is non-degenerate.

Finally, recall that for every finitely generated group and every $n \geq 5$ there are closed connected spin manifolds M of dimension n with $\pi_1(M) = \Gamma$ and which admit a Riemannian metric of positive scalar curvature, compare e.g. [**72**, Example 3.5].

Combined, this gives plenty of examples, e.g. with fundamental groups as in Example 13.8 with infinitely many components in the moduli space of Riemannian metrics of positive scalar curvature.

COROLLARY 13.10. *Let Γ be a hyperbolic group with finite outer automorphism group and let M be a closed connected spin manifold of dimension $n \geq 5$ with $\pi_1(M) = \Gamma$. Assume that M admits a Riemannian metric of positive scalar curvature. Then the virtual rank of the concordance moduli space and even of the bordism moduli space of is bounded below by the rank of the full delocalized cyclic homology in degrees below n, in the appropriate degrees mod 4:*

$$\mathrm{vrank}(\widetilde{\mathcal{P}^+}(M)) \geq \mathrm{vrank}(\widetilde{\Omega\mathcal{P}^+}(M)) \geq \sum_{k>0,p\in\{0,1\}} \mathrm{rank}(H^{n+1-4k-2p}(\Gamma; F^p_{del}\Gamma)).$$

Note that the condition $|\mathrm{Out}(\Gamma)| < \infty$ is in some sense generic for hyperbolic groups, as discussed above.

PROOF. This is a direct consequence of Theorem 13.9 and Theorem 13.1. □

EXAMPLE 13.11. Let the group Γ be hyperbolic or of polynomial growth. Let M be a closed connected spin manifold of odd dimension $n \geq 5$ with $\pi_1(M) = \Gamma$ which admits a metric of positive scalar curvature.

Define the effective number of conjugacy classes of elements of finite order as

$$d := \begin{cases} |\{[\gamma] \subset \Gamma \mid \mathrm{ord}(\gamma) < \infty\}|; & n \equiv -1 \pmod 4 \\ |\{[\gamma] \subset \Gamma \mid [\gamma] \neq [\gamma^{-1}]\}|; & n \equiv 1 \pmod 4 \end{cases}$$

Then
$$\operatorname{vrank}(\tilde{\mathcal{P}^+}(M)) \geq \operatorname{vrank}(\widetilde{\Omega\mathcal{P}^+}(M)) \geq d.$$

Note that elements of different finite order are never conjugate to each other, so
$$d \geq \begin{cases} |\{k>1 \mid \exists\gamma\in\Gamma \text{ with } \operatorname{ord}(\gamma)=k\}|\,; n\equiv -1 \pmod 4 \\ |\{k>1 \mid \exists\gamma\in\Gamma \text{ with } \operatorname{ord}(\gamma)=k \text{ and } [\gamma]\neq[\gamma^{-1}]\}|\,; \quad n\equiv 1 \pmod 4). \end{cases}$$

This way, the example strengthens for our special class of groups Γ the main result [**81**, Theorem 5.8], where $\operatorname{vrank}(\tilde{\mathcal{P}^+}(M))$ is bounded below by the number of orders of elements of finite order in Γ (note that the assertion there for $\dim(M)\equiv 1 \pmod 4$ is not correct, one has to rule out γ with $[\gamma]=[\gamma^{-1}]$, as corrected also in [**84**]).

REMARK 13.12. We started with a closed n-dimensional connected spin manifold M with fundamental group Γ. Our approach here to distinguish elements of $\pi_0(\operatorname{Riem}^+(M)/\operatorname{Diffeo}(M))$ is as follows: we map $\pi_0(\operatorname{Riem}^+(M))$ to the bordism group $\operatorname{Pos}_n^{\mathrm{spin}}(M)$ and then use the delocalized higher rho number as an invariant defined on $\operatorname{Pos}_n^{\mathrm{spin}}(M)$ which descends to the quotient and has infinite image on it.

Here, one might wonder why one does not map to the set $\mathcal{P}^+(M)$ of concordance classes of metrics of positive scalar curvature on M and then use the identification $\mathcal{P}^+(M)\cong R_{n+1}^{\mathrm{spin}}(\Gamma)$ and then a not necessarily delocalized higher rho number obtained as composition
$$\pi_0(\operatorname{Riem}^+(M))\to\mathcal{P}^+(M)\to \mathrm{R}_{n+1}^{\mathrm{spin}}(\Gamma)\xrightarrow{\operatorname{Ind}^\Gamma} K_{n+1}(C^*\Gamma)\xrightarrow{\varrho}\mathbb{C}.$$

The reason is that the identification $\mathrm{R}_{n+1}^{\mathrm{spin}}(\Gamma)\cong\mathcal{P}^+(M)$ is too non-canonical and therefore not controlled enough: in reality, we can only change a metric g_0 to a second metric g_1 by "adding" an element of $R_{n+1}^{\mathrm{spin}}(M)$, so the map gives a well-defined higher rho invariant only for the difference of two metrics, not for an individual one. There seems no way to control the action of the diffeomorphism group on this. The identification of $\mathcal{P}^+(M)$ with $\mathrm{R}_{n+1}^{\mathrm{spin}}(\Gamma)$ depends on the choice of a base point g_0 in $\mathcal{P}^+(M)$. We would like to choose g_0 invariant under the action of $\operatorname{Diffeo}(M)$, which is impossible except for degenerate cases.

Hence, $\operatorname{Pos}_n^{\mathrm{spin}}(M)$ is much more convenient, because a metric g_0 on M defines canonically an element here, and therefore we can assign in a canonical way the delocalized higher rho invariants. We still have to fight with the fact that in general we can't compute the rho invariant explicitly, but have only ways to control how it changes, which leads to the conditions in the meta Theorem 13.1.

REMARK 13.13. The infinitely many classes in $\pi_0(\operatorname{Riem}^+(M)/\operatorname{Diffeo}(M))$ detected in [**81**] are of the same type as ours. However, Xie and Yu focus on degree 0 cyclic cocycles, for which the extension to dense and holomorphically closed subalgebras seems a bit easier. Therefore weaker conditions on the group than ours, namely "finite embeddability" suffice in [**81**]. To make the cocycles invariant, implicitly some averaging process is used there: instead of using the cocycle concentrated on one conjugacy class, it is transported to all conjugacy classes obtained from the given one by the action of the diffeomorphism group (or rather: any finitely generated subgroup). Because of this, the condition of "finite embeddability" has to be strengthened a bit in [**81**].

Prior to [**81**], there is [**64**] which proves that the moduli space is infinite whenever the fundamental group Γ of M contains an element $e\neq\gamma\in\Gamma$ of finite order, and $\dim(M) \equiv 3 \pmod 4$. This uses the L^2-rho invariant which, again, is just a cyclic cocycle of degree 0. Here, this even extends to a C^*-completion of $\mathbb{C}\Gamma$ (the completion associated to the direct sum of the regular representation and the trivial representation). For K-amenable Γ, one can work with the reduced completion, and the result of [**64**] is just a special case of the work done here.

For arbitrary groups, one should observe that it is not difficult to work out a generalized version of what we develop in the paper at hand which works for the relevant completions of the group C^*-algebra. There are two approaches:

- Instead of working on the universal covering, one can stick systematically to operators acting on the Mishchenko bundle. Then one can use arbitrary C^*-completions, but leaves a bit the very direct geometric description given here (operators on $\widetilde{M}$ equivariant for the action of Γ)
- One can also work simultaneously with several coverings and "coupled" operators on them (for the case at hand used in [**64**], it suffices to work with operators just on $\widetilde{M}$ and M). This allows to stick to the classical equivariant description given here, the details of such an approach are explained in [**71**, Section 3].

CHAPTER 14

Geometric applications of relative cohomology

Assume that Γ has property (RD) and $H^*_{pol}(\Gamma) \to H^*(\Gamma)$ is an isomorphism. Then from Proposition 8.14 and Corollary 8.6, we obtain the following commutative diagram

(14.1)

$$\begin{array}{ccccccccc} H^*(M \to B\Gamma) & \longrightarrow & H^*(B\Gamma) & \longrightarrow & H^*(M) & \xrightarrow{\delta} & H^{*+1}(M \to B\Gamma) & \longrightarrow \\ \downarrow & & \downarrow & & \downarrow & & \downarrow & \\ HC^{*-1}(\mathfrak{m}) & \longrightarrow & HC^*(\Psi^{-\infty}_{\mathcal{A}\Gamma}(\widetilde{M})) & \xrightarrow{\partial^{HC}_\Gamma} & HC^*(\pi^*) & \xrightarrow{(1,\sigma_{pr})^*} & HC^*(\mathfrak{m}) & \longrightarrow \end{array}$$

and then the pairing of Theorem 8.24 fits in a map from the pair sequence in cohomology

$$\cdots \to H^*(M \to B\Gamma) \xrightarrow{\iota^*} H^*(B\Gamma) \xrightarrow{u^*} H^*(M) \xrightarrow{\partial} H^{*+1}(M \to B\Gamma) \to \cdots$$

to the dual of the Higson-Roe analytic surgery exact sequence

$$\cdots \to \mathrm{S}^\Gamma_{*-1}(\widetilde{M})' \xrightarrow{s'} K_*(C^*_{red}\Gamma)' \xrightarrow{\mu'_M} K_*(M)' \xrightarrow{c'} \mathrm{S}^\Gamma_*(\widetilde{M})' \to \cdots$$

Here, the map $H^*(M) \to K_*(M)'$ –or equivalently the pairing between $K_*(M)$ and $H^*(M)$– is given by the Chern character, and the pairing between $H^*(\Gamma)$ and $K_*(C^*_{red}\Gamma)$ is the Connes-Moscovici pairing.

In other words, if $\tilde{\alpha} \in H^*(M \to B\Gamma)$ is mapped to $\alpha \in H^*(B\Gamma)$ in the pair sequence and if $x \in K_*(C^*_{red}\Gamma)$, then

$$\langle s(x), \tilde{\alpha} \rangle = \langle x, \alpha \rangle. \tag{14.2}$$

Similarly, if $y \in \mathrm{S}^\Gamma_*(\widetilde{M})$ and $\beta \in H^*(M)$, then

$$\langle c(y), \beta \rangle = \langle y, \partial(\beta) \rangle. \tag{14.3}$$

REMARK 14.1. As we can deduce from diagram (11.7), the pairings and maps above are natural for the action of the diffeomorphism group of M. In particular, Diffeo(M) preserves the kernel of $u^*\colon H^*(B\Gamma) \to H^*(M)$, and of course also the integral cohomology lattice inside this kernel.

If this kernel is 1-dimensional in degree j, then the integral lattice is infinite cyclic and its automorphism group is $\mathbb{Z}/2$. Therefore, a finite index subgroup of Diffeo(M) then acts trivially on $\ker(u^*)\colon H^j(B\Gamma) \to H^j(M)$.

PROPOSITION 14.2. *Let M be a spin manifold of* $\dim(M) = n \geq 5$ *with a metric g of positive scalar curvature. Let $u\colon M \to B\Gamma$ be the classifying map for a universal covering of M; in particular, $\Gamma \cong \pi_1(M)$ and assume that Γ has property (RD) and $H^*_{pol}(\Gamma) \to H^*(\Gamma)$ is an isomorphism. For example, Γ could be hyperbolic or of polynomial growth.*

Assume that the subspace $H \subset H^*(M \to B\Gamma;\mathbb{Q})$ *is pointwise fixed by the action of a finite index subgroup* $U \subset \mathrm{Diffeo}^{\mathrm{spin}}(M)$ *and is mapped bijectively to* $H' \subset H^*(B\Gamma;\mathbb{Q})$ *under the restriction map* $r\colon H^*(M \to B\Gamma;\mathbb{Q}) \to H^*(B\Gamma;\mathbb{Q})$. *Assume* $\dim(H) = d > 0$.

Then the virtual rank of the concordance moduli space and even of the bordism moduli space is bounded below by k:

$$\mathrm{vrank}(\widetilde{\mathcal{P}^+(M)}) \geq \mathrm{vrank}(\widetilde{\Omega\mathcal{P}^+(M)}) \geq d.$$

PROOF. Because Γ has property (RD) and polynomial cohomology, it also satisfies the strong Novikov conjecture. By [**72**, Theorem 1.5] and its proof (which is based on [**84**]), there is a subgroup $B \subset KO_{n+1}(C^*_{\mathbb{R},red}\Gamma)$ and for all $\beta \in B$ there are metrics g_β of positive scalar curvature on M such that

$$\varrho(g_\beta) = \varrho(g) + \iota(\beta) \quad \forall \beta \in B$$

and such that the pairing between H' and B is non-degenerate. Then for each $\alpha \in H$ and $\beta \in B$

$$\langle \varrho(g_\beta), \alpha\rangle = \langle \varrho(g), \alpha\rangle + \langle \iota(\beta), \alpha\rangle = \langle \varrho(g), \alpha\rangle + \langle \beta, r(\alpha)\rangle$$

As the pairing between H' and B is non-degenerate and $\dim(H') = d$, this implies that the $(g_\beta)_{\beta\in B}$ span an affine subgroup of $\mathrm{Pos}^{\mathrm{spin}}_n(M)$ of rank $\geq d$. Due to the fact that the α are fixed by the action of $U \subset \mathrm{Diffeo}^{\mathrm{spin}}(M)$, as in the proof of Theorem 13.1 this remains true for their images in the coinvariant group $\mathrm{Pos}^{\mathrm{spin}}_n(M)_U$ and the assertion about the virtual rank of the bordism moduli space follows. □

EXAMPLE 14.3. In the situation of Proposition 14.2, assume specifically that

$$\dim(\ker(u^*\colon H^{n+1-4k_j}(B\Gamma) \to H^{n+1-4k_j}(M))) = 1 \quad \text{for } k_1 < k_2 < \cdots < k_d.$$

Additionally, assume that the degree $4k_j$-component of the $\hat{A}$-genus of M vanishes: $0 = \hat{A}(M)_{[4k_j]} \in H^{4j}(M;\mathbb{Q})$ for $j = 1, \ldots, d$. This is the case in particular if also $H^{n+1-4k_j}(M \to B\Gamma;\mathbb{Q})$ is 1-dimensional.

Then the virtual rank of the concordance moduli space and even of the bordism moduli space of is bounded below by k:

$$\mathrm{vrank}(\tilde{\mathcal{P}^+}(M)) \geq \mathrm{vrank}(\widetilde{\Omega\mathcal{P}^+(M)}) \geq d.$$

PROOF. For a diffeomorphism Ψ of M we have the commutative diagram of pair sequences in cohomology
(14.4)

$$\begin{array}{ccccccccc}
\longrightarrow & H^{*-1}(M) & \xrightarrow{\delta} & H^*(M \to B\Gamma) & \longrightarrow & H^*(B\Gamma) & \xrightarrow{u^*} & H^*(M) & \longrightarrow \\
& \downarrow \psi^* & & \downarrow (\psi, B\alpha_\psi)^* & & \downarrow B\alpha_\psi * & & \downarrow \psi_* & \\
\longrightarrow & H^{*-1}(M) & \xrightarrow{\delta} & H^*(M \to B\Gamma) & \longrightarrow & H^*(B\Gamma) & \xrightarrow{u^*} & H^*(M) & \longrightarrow
\end{array}$$

In particular, the kernel of u^* inside $H^*(B\Gamma)$ is preserved by every diffeomorphism ψ of M. Because in degree $n + 1 - 4k_j$ this kernel is 1-dimensional by assumption for $j = 1, \ldots, d$, the subgroup U of spin structure preserving diffeomorphism which acts trivially on all these kernels has finite index in $\mathrm{Diffeo}(M)$ (the argument is parallel to the one of Section 12). Let $H \subset H^*(M \to B\Gamma)$ be the inverse image of $\bigoplus_{j=1}^d \ker(u^*\colon H^{n+1-4k_j}(B\Gamma) \to H^{n+1-4k_j}(M))$. By the exactness of the pair sequence, $\dim(H) \geq d$. By commutativity of (14.4), if $\psi \in U$ and $\alpha \in H$ then

$$\psi^*\alpha = \alpha + \delta(\lambda_\alpha) \qquad \text{for some } \lambda_\alpha \in H^{n-4j}(M)$$

We then get for an arbitrary metric g' of positive scalar curvature on M, by using (14.3) and (11.5), that

$$\langle \varrho(\psi^* g'), \alpha\rangle = \langle \varrho(g'), \psi^*\alpha\rangle = \langle \varrho(g'), \alpha\rangle + \langle \varrho(g'), \delta(\lambda_\alpha)\rangle = \langle \varrho(g'), \alpha\rangle + \langle c(\varrho(g')), \lambda_\alpha\rangle.$$

By the commutativity of the third square in (10.2), $c(\varrho(g')) = \beta([M]) \in K_n(M)$ is the K-homology fundamental class. It is a standard fact that the K-homology fundamental class is mapped to the dual of the total $\hat{A}$-genus under the homological Pontryagin character. Consequently, for the relevant pairing we get

$$\langle c(\varrho(g')), \lambda_\alpha\rangle = \langle \beta([M]), \lambda_\alpha\rangle = \int_M \hat{A}(M) \cup \lambda_\alpha = 0.$$

Here, we use our assumption that $\hat{A}(M)_{[4k_j]} = 0$ and $\lambda_\alpha \in H^{n-4k_j}(M)$. Therefore we get finally

$$\varrho_{\tilde{\alpha}}(g') = \varrho_{\tilde{\alpha}}(\psi^* g') \qquad \forall g \in U.$$

It follows that we have a subspace $H \subset H^*(M \to B\Gamma; \mathbb{Q})$ with $\dim(H) \geq d$ and such that the pairing of $\varrho(g')$ with elements of H is fixed under the action of $U \subset \mathrm{Diffeo}^{\mathrm{spin}}(M)$ on $\mathrm{Riem}^+(M)$. Exactly as in the proof of Proposition 14.2 this implies the desired lower bound on the virtual rank of the bordism moduli space.

Let us finally prove that indeed $\hat{A}(M)_{[k_j]} = 0$ if also $\dim(H^{n+1-4k_j}(M \to B\Gamma)) = 1$. This is equivalent to the vanishing of the Poincaré dual $\mathrm{Ph}([M])_{n-4j} \in H_{n-4j}(M; \mathbb{Q})$. Because M has positive scalar curvature and Γ satisfies the strong Novikov conjecture, the image of the K-homology fundamental class of M in $K_*(B\Gamma) \otimes \mathbb{Q}$ vanishes. Dually, this means that $\mathrm{Ph}([M])_{n-4k_j}$ has trivial pairing with the image of the cohomology of $B\Gamma$.

Now consider the long exact cohomology sequence

$$\to H^{n-4k_j}(B\Gamma) \xrightarrow{u^*} H^{n-4k_j}(M) \xrightarrow{\delta} H^{n+1-4k_j}(M \to B\Gamma)$$
$$\to H^{n+1-4k_j}(B\Gamma) \xrightarrow{H^{n+1-4k_j}(u)} H^{n+1-4k_j}(M) \to$$

By assumption, $\dim\ker(H^{n+1-4k_j}(u)) = 1 = \dim H^{n+1-4k_j}(M \to B\Gamma)$. Therefore, $H^{n+1-4k_j}(M \to B\Gamma) \to H^{n+1-4k_j}(B\Gamma)$ is injective, δ is the zero map and $H^{n-4k_j}(u)$ is surjective. By the reasoning above this means that $\mathrm{Ph}([M])_{n-4k_j}$ pairs trivially with $H^{n-4k_j}(M)$, i.e. $\mathrm{Ph}([M])_{n-4k_j} = 0$ and consequently also its Poincaré dual $\hat{A}(M)_{[4k_j]} = 0$. □

REMARK 14.4. The (higher) rho invariants based on group cohomology, as developed in Section 12, ultimately relies on the existence of torsion in the fundamental group. This is explicitly explained in [**84**], but was evident from the beginning of the use of rho invariants to distinguish metrics of positive scalar curvature, see for example [**12, 64**].

On the other hand, the use of the relative cohomology $H^*(M \to B\Gamma)$ as in the present section is *independent on the existence of torsion in* Γ.

Notice that if Γ is torsion free and the Baum-Connes map is an isomorphism then the "lower" numeric rho invariants, that is, the Atiyah-Patodi-Singer rho invariant, the Cheeger-Gromov rho-invariant and Lott's delocalized eta invariant, are all equal to zero and therefore of no use in distinguishing metrics of positive scalar curvature. See [**41**], [**42**], [**63**], [**33**], [**8**].

REMARK 14.5. The metrics which are distinguished in Proposition 14.2 not only lie in different components of the moduli space $\mathrm{Riem}^+(M)/\mathrm{Diffeo}(M)$, they are also (by the way they are detected) not bordant to each other (with reference map to M). This distinguishes them from families of metrics obtained by Carr [**14**], which are shown to be bordant in [**72**, Remark 3.4]. Moreover, Carr gives infinitely many non-concordant metrics of positive scalar curvature on any spin manifold M of dimension $4d-1$. But these are only known to represent infinitely many different components of the moduli space $\mathrm{Riem}^+(M)/\mathrm{Diffeo}(M)$ for very special M, e.g. if M is a homology sphere or if M is stably parallelizable, compare [**29**, Theorem 1.11].

It is clear that there are plenty of situations where Proposition 14.2 can be applied. The following example gives a very general method to construct some.

EXAMPLE 14.6. Let Γ be a discrete group with a classifying space $B\Gamma$ with finite k-skeleton $B\Gamma^{(k)}$ and assume that $H^k(\Gamma;\mathbb{Q}) \neq \{0\}$ for some $k \geq 2$.

Construct a sub-CW-complex X of $B\Gamma^{(k)}$ as follows. Start with $Y = B\Gamma^{(k)}$ and leave out one k-cell after the other to obtain $Y \supset Y_1 \supset Y_2 \supset \cdots$. We get a corresponding filtration of the k-cellular chain group

$$C_k(B\Gamma)\otimes\mathbb{Q} = C_k(Y)\otimes\mathbb{Q} \supset C_k(Y_1)\otimes\mathbb{Q} \supset C_k(Y_2)\otimes\mathbb{Q}\cdots \supset \{0\}.$$

We have a surjective map $\ker(\partial_k) \to H_k(B\Gamma) = \ker(\partial_k)/\operatorname{im}(\partial_{k+1})$. Restricting ∂_k to the decreasing chain of subspace $C_k(Y_j)\otimes\mathbb{Q}$ we get a decreasing sequence of kernels starting with $\ker(\partial_k)$ and ending in $\{0\}$. Therefore, there will be a subspace Y_j such that $\ker(\partial_k|_{C_k(Y_j)}) \to H_k(B\Gamma)$ has codimension 1. Set $X := Y_j$. Because X is an k-dimensional CW-complex, $H_k(X) = \ker(\partial_k|_{C_k(X)})$ which maps to $H_k(B\Gamma)$ with 1-dimensional cokernel, whereas $H_j(X) \to H_j(B\Gamma)$ is an isomorphism for $j < k$ and also an isomorphism on π_1, as $X^{(k-1)} = B\Gamma^{(k-1)}$.

Choose now $n \geq 2k$ and embed X into $\mathbb{R}^{2n+1}$. Take a thickening W of the embedding: a compact "tubular neighbourhood" manifold with boundary such that the inclusion $X \to W$ is a homotopy equivalence. Set $M := \partial W$. By general position, the inclusion $M \to W$ is an $n-k \geq k$-equivalence (an isomorphism on π_j for $j < n-k$ and an epimorphism for $j = n-k$). In particular, $\pi_1(M) = \Gamma$ and

$$H_k(M) \to H_k(W) = H_k(X) \to H_k(B\Gamma)$$

has still 1-dimensional cokernel.

By construction, W is a spin null-bordism of M with reference map to $B\Gamma$ or, after cutting out a disk, a $B\Gamma$ spin bordism from S^n to M. By the Gromov-Lawson surgery theory [**28**] (compare e.g. [**72**, Section 3] and [**24**]), M admits a Riemannian metric of positive scalar curvature.

Let us give another example of slightly different nature.

EXAMPLE 14.7. Let us take Γ equal to $\mathbb{Z}^k$ for $k \geq 3$, or more generally Γ finitely presented with

$$\dim(H_k(\Gamma)) = 1.$$

Observe that $B\mathbb{Z}^k$ is realized by the k-dimensional torus $\mathbb{T}^k$.

Let us construct a suitable manifold which fits with the situation in Proposition 14.2. Let us pick a finite presentation $\Gamma = \langle x_1, \dots, \xi_g \mid r_1, \dots, r_h\rangle$. Take then the wedge of g circles and, for each relation r_l, attach a two cell. Denote by X this 2-dimensional CW-complex.

Choose $n = k - 1 + 4j$ for some $j \in \mathbb{N}_{>0}$ and embed X into $\mathbb{R}^{n+1}$, which is possible if $n \geq 5$. Now, consider a tubular neighbourhood $\mathcal{N}$ of X, hence its boundary $M := \partial\mathcal{N}$ is an n-dimensional spin manifold with fundamental group Γ. Moreover, since $\mathcal{N}$ is a spin null-bordism for M with reference map to $B\Gamma$, it follows that M admits a metric with positive scalar curvature. The manifold W has a trivial tangent bundle because its interior is an open subset of $\mathbb{R}^{n+1}$, and therefore the tangent bundle of M is stably parallelizable and $\hat{A}(M) = 1$ is trivial.

The canonical map $M \to B\Gamma$ factors as $M \to X \to B\Gamma$. Now X is a 2-dimensional CW-complex and therefore $H_k(X) = 0$, as $k \geq 3$.

If Γ satisfies the technical conditions that it has property (RD) and that $H^*_{pol}(\Gamma) \to H^*(\Gamma)$ is an isomorphism, which is certainly the case for $\Gamma = \mathbb{Z}^k$, then all conditions of Proposition 14.2 are fulfilled and the moduli space

$$\mathrm{Riem}^+(M)/\,\mathrm{Diffeo}(M)$$

has infinitely many components.

CHAPTER 15

Geometric applications: A topological approach

There is a different approach to the definition of higher rho numbers associated to elements in $H^*(M \to B\Gamma)$, based heavily on the Baum-Connes map being an isomorphism and due to Weinberger, Xie and Yu, see [**79**, Section 7]. In this Section we want to compare the two approaches and also elaborate on how the applications we have given involving the diffeomorphism group can be extended to the context treated in [**79**, Section 7].

The approach of this paper for understanding the Higson-Roe analytic exact sequence via non-commutative cohomology is a classical one, employed initially to get explicit understanding of the term $K_*(C^*_{red}\Gamma)$ in that sequence. The main result of this paper is to show that this approach extends systematically to the whole sequence, in particular to the structure group $\mathrm{S}^\Gamma_*(\widetilde{M})$.

We give specific examples where this is particularly well controlled, in particular for groups Γ which are hyperbolic or of polynomial growth, and we explain specific geometric applications of our approach (which of course work best when the homology side is well understood, e.g. again for groups Γ which are hyperbolic or of polynomial growth).

To understand $K_*(C^*_{red}\Gamma)$ there is of course the Baum-Connes conjecture and the Baum-Connes assembly map $K^\Gamma_*(\underline{E}\Gamma) \xrightarrow{\mu} K_*(C^*_{red}\Gamma)$. It is pretty clear that there should be a natural map from the long exact pair sequence in equivariant K-homology to the whole Higson-Roe analytic exact sequence:

(15.1)

$$\begin{array}{ccccccccc}
\longrightarrow & K^\Gamma_*(\widetilde{M}) & \longrightarrow & K^\Gamma_*(\underline{E}\Gamma) & \longrightarrow & K^\Gamma_*(\widetilde{M} \to \underline{E}\Gamma) & \longrightarrow & K^\Gamma_{*-1}(\tilde{M}) & \longrightarrow \\
 & \big\downarrow{\scriptstyle =} & & \big\downarrow{\scriptstyle \mu} & & \big\downarrow{\scriptstyle \Lambda} & & \big\downarrow{\scriptstyle =} & \\
\longrightarrow & K^\Gamma_*(\widetilde{M}) & \longrightarrow & K_*(C^*_{red}\Gamma) & \longrightarrow & \mathrm{S}^\Gamma_{*-1}(\widetilde{M}) & \longrightarrow & K^\Gamma_{*-1}(\widetilde{M}) & \longrightarrow
\end{array}$$

This is (briefly) stated in [**79**, diagram (13)]. Using the diagram (15.1), the 5-lemma immediately gives the following corollary.

COROLLARY 15.1. *Assume that the Baum-Connes assembly map $\mu\colon K_*(\underline{E}\Gamma) \to K_*(C^*_{red}\Gamma)$ is an isomorphism and that a natural commutative diagram such as* (15.1) *has been constructed. Then we get a natural isomorphism*

$$\Lambda\colon K^\Gamma_*(\widetilde{M} \to \underline{E}\Gamma) \xrightarrow{\cong} \mathrm{S}^\Gamma_{*-1}(\widetilde{M}).$$

So, in this case, we have full understanding of the structure group $\mathrm{S}^\Gamma_*(\widetilde{M})$.

Note that the Baum-Connes conjecture predicts that $\mu\colon K^\Gamma_*(\underline{E}\Gamma) \to K_*(C^*_{red}\Gamma)$ is an isomorphism for every group Γ. This conjecture is known for large classes of groups, in particular for Gromov hyperbolic groups [**61**] and for groups of polynomial growth, which are our main examples. No counterexamples to the conjecture are known to this date.

Let us have a closer look at the term $K_*^\Gamma(\widetilde{M} \to \underline{E}\Gamma)$ which conjecturally computes $\mathrm{S}_{*-1}^\Gamma(\widetilde{M})$.

PROPOSITION 15.2. *The Chern character provides a natural isomorphism*

$$\mathrm{Ch}\colon K_*^\Gamma(\widetilde{M} \to \underline{E}\Gamma) \otimes \mathbb{C} \xrightarrow{\cong} \bigoplus_{k\in\mathbb{Z}} H_{*+2k}(M \to B\Gamma; \mathbb{C}) \oplus \bigoplus_{k\in\mathbb{Z}} H_{*+2k}(\Gamma, F_{del}\Gamma).$$

Here, the Γ-module $F_{del}\Gamma$ is the delocalized summand of $F\Gamma$ of Chapter 12, i.e. $F_{del}\Gamma = \{f\colon \Gamma_{fin} \setminus \{e\} \to \mathbb{C} \mid |\mathrm{supp}(f)| < \infty\}$ with conjugation action, where Γ_{fin} is the set of elements $e \neq \gamma \in \Gamma$ of finite order. We have the obvious decomposition $F\Gamma = F_{del}\Gamma \oplus \mathbb{C}$ with $\mathbb{C} = \{f\colon \{e\} \to \mathbb{C}\}$ the trivial Γ-representation.

As in Section 12, we have the splitting over all conjugacy classes $[\gamma]$ of elements of finite order

$$H_*(\Gamma; F_{del}\Gamma) \cong \bigoplus_{[\gamma], \gamma\neq e} H_*(\Gamma_\gamma; \mathbb{C}).$$

All this is compatible with the maps of the pair sequence in equivariant K-homology and the natural Chern character isomorphism

$$K_*^\Gamma(\tilde{M}) \otimes \mathbb{C} \xrightarrow{\cong} \bigoplus_{k\in\mathbb{Z}} H_{*+2k}(M; \mathbb{C});$$

$$K_*^\Gamma(\underline{E}\Gamma) \otimes \mathbb{C} \xrightarrow{\cong} \bigoplus_{k\in\mathbb{Z}} H_{*+2k}(\Gamma; \mathbb{C}) \oplus \bigoplus_{k\in\mathbb{Z}} H_{*+2k}(\Gamma; F_{del}\Gamma),$$

where of course $H_*(\Gamma; \mathbb{C}) = H_*(B\Gamma; \mathbb{C})$.

PROOF. This follows directly by comparison with the Chern character for the pair sequence

$$\cdots \to K_*(M) \to K_*(B\Gamma) \to K_*(M \to B\Gamma) \to \cdots$$

mapping to the Chern character of the pair sequence

$$\cdots \to K_*^\Gamma(\widetilde{M}) \to K_*^\Gamma(\underline{E}\Gamma) \to K_*^\Gamma(\tilde{M} \to \underline{E}\Gamma) \to \cdots$$

via the natural isomorphisms $K_*^\Gamma(\tilde{X}) \to K_*(X)$ for a space $\tilde{X}$ with free Γ-action and with quotient $X = \tilde{X}/\Gamma$. In the terms for $\underline{E}\Gamma$ then $H_*(B\Gamma; \mathbb{C})$ injects to $H_*(\Gamma; F\Gamma)$ as the summand localized at the identity, while in the terms for $\widetilde{M}$, $H_*(M; \mathbb{C})$ is the full image of the Chern character.

Consequently, $H_*(M \to B\Gamma; \mathbb{C})$ injects as a direct summand into the image of the Chern character isomorphism for $(\widetilde{M} \to \underline{E}\Gamma)$ with remaining summand coming from $\underline{E}\Gamma$, where it is precisely the delocalized homology $H_*(\Gamma; F_{del}\Gamma)$. □

COROLLARY 15.3. *Assume that the Baum-Connes map is an isomorphism. Then this induces natural pairings of $S_{*-1}^\Gamma(\widetilde{M})$ with $H_{*+2k}(M \to B\Gamma; \mathbb{C})$ and with $H_{*+2k}(\Gamma; F_{del}\Gamma)$ for all $k \in \mathbb{Z}$ which are compatible with the pairings of $H_{*+2k}(\Gamma; F_{del}\Gamma)$ and $H^*(\Gamma; \mathbb{C})$ with $K_*(C_{red}^*\Gamma)$ and the pairing of $H^*(M; \mathbb{C})$ with $K_*(M)$.*

PROOF. This just a consequence of the map to homology, where we can then pair with cohomology as usual. □

REMARK 15.4. Throughout, we can pass to real K-theory and obtain a diagram corresponding to (15.1) with the real K-homology and structure group throughout.

The Chern character isomorphism of Proposition 15.2 then has the analog for real K-homology using the splitting into even and odd part of $F_{del}\Gamma$ as in Chapter 12, in particular in (12.10).

We shall now briefly treat the action of the diffeomorphism group in this context, using the ideas we have explained in the previous sections. Naturality of the constructions implies that we have the expected action of $\mathrm{Diffeo}(M)$ on the whole diagram (15.1). Under the Chern character isomorphism of Proposition 15.2 the action on the group cohomology terms factors through the homomorphism $\mathrm{Diffeo}(M) \to \mathrm{Out}(\Gamma)$. The methods we have developed in Chapter 12 and Chapter 14 apply without change here and we obtain the following corollary:

COROLLARY 15.5. *Assume that Γ satisfies the Baum-Connes conjecture so that there exists a Baum-Connes isomorphism $K^{\Gamma}_{*+1}(\widetilde{M} \to \underline{E}\Gamma) \xrightarrow{\cong} \mathrm{S}^{\Gamma}_{*}(\widetilde{M})$ (coming from the commutative diagram (15.1)). Assume that M is a spin manifold of dimension $n \geq 5$ with positive scalar curvature. Assume that a class $0 \neq \alpha \in H_{n+1-4j}(\widetilde{M} \to \underline{E}\Gamma; \mathbb{C})$ is preserved by the action of $\mathrm{Diffeo}(M)$ for some $j > 0$. Then the moduli space $\mathrm{Riem}^{+}(M)/\mathrm{Diffeo}(M)$ has infinitely many components.*

For example, we obtain such a homology class α whenever a summand

$$H_{n+1-4j}(M \to B\Gamma)$$

or a delocalized group homology summand as in (12.10) of degree $n+1-2p-4j$ for some $j > 0$ is 1-dimensional, as in Chapters 12 or 14.

REMARK 15.6. Assume, more generally than in Corollary 15.5, that the group Γ satisfies that the Baum-Connes assembly map $\mu\colon K^{\Gamma}_{*}(\underline{E}\Gamma) \to K_*(C^*_{red}\Gamma)$ is rationally split injective with a split $\alpha\colon K_*(C^*_{red}\Gamma) \otimes \mathbb{Q} \to K^{\Gamma}_{*}(\underline{E}\Gamma) \otimes \mathbb{Q}$ which is natural for the obvious action of $\mathrm{Out}(\Gamma)$. Then we can state and prove the analogue of Corollary 15.5.

PROOF. By routine diagram chases, as spelled out in [**79**, Section 7], the assumptions give a natural split injection of the pair sequence into the Higson-Roe exact analytic sequence (where we omit the tensor factor $\otimes\mathbb{Q}$ for better readability):

$$\begin{array}{ccccccccc}
\longrightarrow & K^{\Gamma}_{n}(\widetilde{M}) & \longrightarrow & K^{\Gamma}_{n}(\underline{E}\Gamma) & \longrightarrow & K^{\Gamma}_{n}(\widetilde{M} \to \underline{E}\Gamma) & \longrightarrow \\
& \big\downarrow{=} & & \big\downarrow & & \big\downarrow & \\
\longrightarrow & K^{\Gamma}_{n}(\widetilde{M}) & \longrightarrow & K^{\Gamma}_{n}(\underline{E}\Gamma) \oplus \mathcal{E} & \longrightarrow & K^{\Gamma}_{n}(\tilde{M} \to \underline{E}\Gamma) \oplus \mathcal{E} & \longrightarrow \\
& \big\downarrow{=} & & \cong\big\downarrow{\mu\oplus\iota} & & \cong\big\downarrow{\Lambda\oplus\iota} & \\
\longrightarrow & K^{\Gamma}_{n}(\widetilde{M}) & \longrightarrow & K_n(C^*_{red}\Gamma) & \longrightarrow & \mathrm{S}^{\Gamma}_{n-1}(\widetilde{M}) & \longrightarrow
\end{array}$$

Here, $\mathcal{E}$ is the natural summand complementary to $\mu(K^{\Gamma}_{n}(\underline{E}\Gamma) \otimes \mathbb{Q})$ of $K_n(C^*_{red}\Gamma) \otimes \mathbb{Q}$ determined by the split α, which by a diagram chase then also becomes the natural summand complementary to $\Lambda(K^{\Gamma}_{n}(\tilde{M} \to \underline{E}\Gamma) \otimes \mathbb{Q})$ in $\mathrm{S}^{\Gamma}(\tilde{M})$.

The same arguments as in the case where the Baum-Connes map is an isomorphism can then be used, ignoring throughout the (unknown) summand $\mathcal{E}$. □

REMARK 15.7. Using Corollary 15.5, if the Baum-Connes map is an isomorphism or more generally, if it is split injective, we can define via this topological route for each cohomology class $[\tau] \in H^{*+2k}(M \to B\Gamma; \mathbb{C})$ or $[\tau] \in H^{*+2k}(\Gamma; F_{del}\Gamma)$ topologically defined numeric rho invariants (say, of a positive scalar curvature metric g on a spin manifold M via its invertible Dirac operator $\tilde{D}$)

$$\varrho_\tau^{BC}(g) := \langle \mathrm{Ch}(\Lambda^{-1}(\varrho(g))), [\tau] \rangle$$

(if the Baum-Connes map only has a split α, we have to replace Λ^{-1} by this split α). In [**79**], $\alpha(\varrho(g))$ is called $\varrho^{Nov}(g)$. The results of Corollary 15.5 apply equivalently to these numeric rho invariants.

A priori, it is not clear if these numbers coincide with the numeric rho invariants we defined earlier in Definitions 7.25 and 8.27. There, the definition is via the Chern character to non-commutative de Rham homology and the paring of the latter with cyclic or relative cohomology.

This depends on the compatibility of our Chern character map with the Baum-Connes version, and depends in particular on the construction of Λ.

Nonetheless, all our computations in Chapters 13 and 14 only compute *differences* of higher rho numbers, which are obtained as values of the pairing with $K_*(C^*_{red}\Gamma)$. For the latter, the compatibility of the Baum-Connes Chern character and the cyclic homology Chern character is well known. Therefore, the *differences* of higher rho numbers we obtain in Chapters 13 and 14 coincide with the differences of the corresponding $\varrho_\tau^{BC}(g)$.

CHAPTER 16

Concluding remarks

We end this article by stressing why we believe that the homological methods developed here (and in contributions by other authors) are of interest, both from a general point of view and also when compared to the more K-theoretic arguments of the last chapter based on the Baum-Connes conjecture:

- the pairings we construct in Chapter 7 are of interest in a very general context; they have been applied here to rho invariants but it is conceivable that other invariants to which these methods apply will be defined in the future;
- we obtain very explicit integral formulas for the higher rho numbers, which a priori might be used for calculations in entirely new situations;
- we give, for hyperbolic groups and groups of polynomial growth, a proof of the geometric applications which does not rely on the proof of the Baum-Connes conjecture;
- our results are a priori more general than just working for groups where the Baum-Connes conjecture is known: whenever one has appropriate control of the homology of a smooth subalgebra $\mathcal{A}\Gamma$, one gets explicit formulas and the desired geometric consequences. There is no reason why this should only work if the Baum-Connes conjecture holds.
- we feel that the (cyclic) homological method has potential to be used in other geometric situations, e.g. when analyzing foliated manifolds for the study of metrics with leafwise positive scalar curvature, where it is not clear how to define the *coarse geometric* target of Λ and Λ itself in that context;
- another geometric situation of relevance for our methods is the one where a reductive Lie group acts properly and isometrically on a manifold. The (cyclic) homological approach has been applied very successfully to primary invariants of Dirac operators and there is in fact a large literature on the subject. For secondary invariants new results through the homological methods adopted here are emerging recently, compare e.g. [**34, 74**].

Bibliography

[1] M. F. Atiyah, *Elliptic operators, discrete groups and von Neumann algebras*, Colloque "Analyse et Topologie" en l'Honneur de Henri Cartan (Orsay, 1974), Astérisque, No. 32-33, Soc. Math. France, Paris, 1976, pp. 43–72. MR0420729 ↑1, 2, 89

[2] M. F. Atiyah, V. K. Patodi, and I. M. Singer, *Spectral asymmetry and Riemannian geometry. II*, Math. Proc. Cambridge Philos. Soc. **78** (1975), no. 3, 405–432, DOI 10.1017/S0305004100051872. MR397798 ↑4

[3] P. Antonini, S. Azzali, and G. Skandalis, *Flat bundles, von Neumann algebras and K-theory with $\mathbb{R}/\mathbb{Z}$-coefficients*, J. K-Theory **13** (2014), no. 2, 275–303, DOI 10.1017/is014001024jkt253. MR3189427 ↑9, 15

[4] P. Antonini, S. Azzali, and G. Skandalis, *Bivariant K-theory with $\mathbb{R}/\mathbb{Z}$-coefficients and rho classes of unitary representations*, J. Funct. Anal. **270** (2016), no. 1, 447–481, DOI 10.1016/j.jfa.2015.06.017. MR3419768 ↑9

[5] S. Azzali and C. Wahl, *Two-cocycle twists and Atiyah-Patodi-Singer index theory*, Math. Proc. Cambridge Philos. Soc. **167** (2019), no. 3, 437–487, DOI 10.1017/s0305004118000427. MR4015646 ↑10

[6] N. Bárcenas and R. Zeidler, *Positive scalar curvature and low-degree group homology*, Ann. K-Theory **3** (2018), no. 3, 565–579, DOI 10.2140/akt.2018.3.565. MR3830202 ↑10, 118

[7] P. Baum and A. Connes, *Chern character for discrete groups*, A fête of topology, Academic Press, Boston, MA, 1988, pp. 163–232, DOI 10.1016/B978-0-12-480440-1.50015-0. MR928402 ↑118

[8] M.-T. Benameur and I. Roy, *The Higson-Roe exact sequence and ℓ^2 eta invariants*, J. Funct. Anal. **268** (2015), no. 4, 974–1031, DOI 10.1016/j.jfa.2014.11.006. MR3296587 ↑9, 132

[9] N. Berline, E. Getzler, and M. Vergne, *Heat kernels and Dirac operators*, Grundlehren Text Editions, Springer-Verlag, Berlin, 2004. Corrected reprint of the 1992 original. MR2273508 ↑57, 58, 59

[10] M. Bestvina and M. Feighn, *Stable actions of groups on real trees*, Invent. Math. **121** (1995), no. 2, 287–321, DOI 10.1007/BF01884300. MR1346208 ↑124

[11] J.-M. Bismut and J. Cheeger, *η-invariants and their adiabatic limits*, J. Amer. Math. Soc. **2** (1989), no. 1, 33–70, DOI 10.2307/1990912. MR966608 ↑58

[12] B. Botvinnik and P. B. Gilkey, *The eta invariant and metrics of positive scalar curvature*, Math. Ann. **302** (1995), no. 3, 507–517, DOI 10.1007/BF01444505. MR1339924 ↑5, 131

[13] D. Burghelea, *The cyclic homology of the group rings*, Comment. Math. Helv. **60** (1985), no. 3, 354–365, DOI 10.1007/BF02567420. MR814144 ↑8, 66, 118, 121

[14] R. Carr, *Construction of manifolds of positive scalar curvature*, Trans. Amer. Math. Soc. **307** (1988), no. 1, 63–74, DOI 10.2307/2000751. MR936805 ↑132

[15] P. Carrillo Rouse, B. L. Wang, and H. Wang, *Topological K-theory for discrete groups and index theory*, Bull. Sci. Math. **185** (2023), Paper No. 103262, 70, DOI 10.1016/j.bulsci.2023.103262. MR4583695 ↑117

[16] S. Chang and S. Weinberger, *On invariants of Hirzebruch and Cheeger-Gromov*, Geom. Topol. **7** (2003), 311–319, DOI 10.2140/gt.2003.7.311. MR1988288 ↑5

[17] J. Cheeger and M. Gromov, *On the characteristic numbers of complete manifolds of bounded curvature and finite volume*, Differential geometry and complex analysis, Springer, Berlin, 1985, pp. 115–154. MR780040 ↑4

[18] X. Chen, J. Wang, Z. Xie, and G. Yu, *Delocalized eta invariants, cyclic cohomology and higher rho invariants*, Ann. K-Theory **8** (2023), no. 4, 531–608, DOI 10.2140/akt.2023.8.531. MR4673948 ↑8, 9, 10, 13, 78

[19] A. Connes, *Noncommutative geometry*, Academic Press, Inc., San Diego, CA, 1994. MR1303779 ↑47, 69

[20] A. Connes and H. Moscovici, *Cyclic cohomology, the Novikov conjecture and hyperbolic groups*, Topology **29** (1990), no. 3, 345–388, DOI 10.1016/0040-9383(90)90003-3. MR1066176 ↑2, 3, 7, 53, 54, 73, 87, 89, 90, 91, 94

[21] A. Van Daele, *K-theory for graded Banach algebras. I*, Quart. J. Math. Oxford Ser. (2) **39** (1988), no. 154, 185–199, DOI 10.1093/qmath/39.2.185. MR947500 ↑

[22] C. Debord and G. Skandalis, *Blow-up constructions for Lie groupoids and a Boutet de Monvel type calculus*, Münster J. Math. **14** (2021), no. 1, 1–40, DOI 10.17879/59019640550. MR4300161 ↑61, 62

[23] R. J. Deeley and M. Goffeng, *Realizing the analytic surgery group of Higson and Roe geometrically part III: higher invariants*, Math. Ann. **366** (2016), no. 3-4, 1513–1559, DOI 10.1007/s00208-016-1365-6. MR3563244 ↑10, 41

[24] J. Ebert and G. Frenck, *The Gromov-Lawson-Chernysh surgery theorem*, Bol. Soc. Mat. Mex. (3) **27** (2021), no. 2, Paper No. 37, 43, DOI 10.1007/s40590-021-00310-w. MR4236952 ↑132

[25] A. Gorokhovsky and J. Lott, *Local index theory over étale groupoids*, J. Reine Angew. Math. **560** (2003), 151–198, DOI 10.1515/crll.2003.054. MR1992804 ↑47, 54, 61

[26] A. Gorokhovsky and J. Lott, *Local index theory over foliation groupoids*, Adv. Math. **204** (2006), no. 2, 413–447, DOI 10.1016/j.aim.2005.05.018. MR2249619 ↑54

[27] M. Gromov, *Hyperbolic groups*, Essays in group theory, Math. Sci. Res. Inst. Publ., vol. 8, Springer, New York, 1987, pp. 75–263, DOI 10.1007/978-1-4613-9586-7_3. MR919829 ↑67, 68, 70, 124

[28] M. Gromov and H. B. Lawson Jr., *The classification of simply connected manifolds of positive scalar curvature*, Ann. of Math. (2) **111** (1980), no. 3, 423–434, DOI 10.2307/1971103. MR577131 ↑132

[29] B. Hanke, T. Schick, and W. Steimle, *The space of metrics of positive scalar curvature*, Publ. Math. Inst. Hautes Études Sci. **120** (2014), 335–367, DOI 10.1007/s10240-014-0062-9. MR3270591 ↑132

[30] N. Higson and J. Roe, *Mapping surgery to analysis. I. Analytic signatures*, *K*-Theory **33** (2005), no. 4, 277–299, DOI 10.1007/s10977-005-1561-8. MR2220522 ↑5, 7

[31] N. Higson and J. Roe, *Mapping surgery to analysis. II. Geometric signatures*, *K*-Theory **33** (2005), no. 4, 301–324, DOI 10.1007/s10977-005-1559-2. MR2220523 ↑7

[32] N. Higson and J. Roe, *Mapping surgery to analysis. III. Exact sequences*, *K*-Theory **33** (2005), no. 4, 325–346, DOI 10.1007/s10977-005-1554-7. MR2220524 ↑7

[33] N. Higson and J. Roe, *K-homology, assembly and rigidity theorems for relative eta invariants*, Pure Appl. Math. Q. **6** (2010), no. 2, Special Issue: In honor of Michael Atiyah and Isadore Singer, 555–601, DOI 10.4310/PAMQ.2010.v6.n2.a11. MR2761858 ↑9, 132

[34] P. Hochs, Y. Song, and X. Tang, *An index theorem for higher orbital integrals*, Math. Ann. **382** (2022), no. 1-2, 169–202, DOI 10.1007/s00208-021-02233-3. MR4377301 ↑139

[35] C. J. Hou, *Spectral invariant subalgebras of reduced groupoid C^*-algebras*, Acta Math. Sin. (Engl. Ser.) **33** (2017), no. 4, 526–544, DOI 10.1007/s10114-016-6264-y. MR3620190 ↑94

[36] R. Ji, C. Ogle, and B. Ramsey, *Relatively hyperbolic groups, rapid decay algebras and a generalization of the Bass conjecture*, J. Noncommut. Geom. **4** (2010), no. 1, 83–124, DOI 10.4171/JNCG/50. With an appendix by Ogle. MR2575390 ↑73

[37] S. A. K. A. John, *Secondary higher invariants and cyclic cohomology for groups of polynomial growth*, J. Noncommut. Geom. **16** (2022), no. 4, 1283–1335, DOI 10.4171/jncg/456. MR4542386 ↑13

[38] P. Jolissaint, *Rapidly decreasing functions in reduced C^*-algebras of groups*, Trans. Amer. Math. Soc. **317** (1990), no. 1, 167–196, DOI 10.2307/2001458. MR943303 ↑73

[39] P. Jolissaint, *K-theory of reduced C^*-algebras and rapidly decreasing functions on groups*, *K*-Theory **2** (1989), no. 6, 723–735, DOI 10.1007/BF00538429. MR1010979 ↑73

[40] M. Karoubi, *Homologie cyclique et K-théorie* (French, with English summary), Astérisque **149** (1987), 147. MR913964 ↑8, 21, 65, 74, 75

[41] N. Keswani, *Relative eta-invariants and C^*-algebra K-theory*, Topology **39** (2000), no. 5, 957–983, DOI 10.1016/S0040-9383(99)00045-2. MR1763959 ↑132

[42] N. Keswani, *Von Neumann eta-invariants and C^*-algebra K-theory*, J. London Math. Soc. (2) **62** (2000), no. 3, 771–783, DOI 10.1112/S0024610700001344. MR1794283 ↑132

[43] R. Lauter, B. Monthubert, and V. Nistor, *Spectral invariance for certain algebras of pseudodifferential operators*, J. Inst. Math. Jussieu **4** (2005), no. 3, 405–442, DOI 10.1017/S1474748005000125. MR2197064 ↑40, 94

[44] É. Leichtnam and P. Piazza, *The b-pseudodifferential calculus on Galois coverings and a higher Atiyah-Patodi-Singer index theorem* (English, with English and French summaries), Mém. Soc. Math. Fr. (N.S.) **68** (1997), iv+121. MR1488084 ↑10, 44, 54, 57, 59, 60

[45] É. Leichtnam and P. Piazza, *Homotopy invariance of twisted higher signatures on manifolds with boundary* (English, with English and French summaries), Bull. Soc. Math. France **127** (1999), no. 2, 307–331. MR1708639 ↑57, 60

[46] E. Leichtnam and P. Piazza, *On higher eta-invariants and metrics of positive scalar curvature*, K-Theory **24** (2001), no. 4, 341–359, DOI 10.1023/A:1014079307698. MR1885126 ↑60

[47] E. Leichtnam and P. Piazza, *Dirac index classes and the noncommutative spectral flow*, J. Funct. Anal. **200** (2003), no. 2, 348–400, DOI 10.1016/S0022-1236(02)00044-7. MR1979016 ↑60

[48] E. Leichtnam and P. Piazza, *Étale groupoids, eta invariants and index theory*, J. Reine Angew. Math. **587** (2005), 169–233, DOI 10.1515/crll.2005.2005.587.169. MR2186978 ↑54

[49] M. Lesch, H. Moscovici, and M. J. Pflaum, *Relative pairing in cyclic cohomology and divisor flows*, J. K-Theory **3** (2009), no. 2, 359–407, DOI 10.1017/is008001021jkt051. MR2496452 ↑27

[50] J.-L. Loday, *Cyclic homology*, Grundlehren der mathematischen Wissenschaften [Fundamental Principles of Mathematical Sciences], vol. 301, Springer-Verlag, Berlin, 1992. Appendix E by María O. Ronco, DOI 10.1007/978-3-662-21739-9. MR1217970 ↑65, 69, 75

[51] J. Lott, *Superconnections and higher index theory*, Geom. Funct. Anal. **2** (1992), no. 4, 421–454, DOI 10.1007/BF01896662. MR1191568 ↑35, 36, 37, 40, 42, 44, 54, 57, 59, 60

[52] J. Lott, *Higher eta-invariants*, K-Theory **6** (1992), no. 3, 191–233, DOI 10.1007/BF00961464. MR1189276 ↑9, 10, 56, 57, 59, 61, 78, 79, 80

[53] J. Lott, *Delocalized L^2-invariants*, J. Funct. Anal. **169** (1999), no. 1, 1–31, DOI 10.1006/jfan.1999.3451. MR1726745 ↑4

[54] J. Lott, *Diffeomorphisms and noncommutative analytic torsion*, Mem. Amer. Math. Soc. **141** (1999), no. 673, viii+56, DOI 10.1090/memo/0673. MR1618772 ↑40, 57

[55] W. Lück, *The type of the classifying space for a family of subgroups*, J. Pure Appl. Algebra **149** (2000), no. 2, 177–203, DOI 10.1016/S0022-4049(98)90173-6. MR1757730 ↑70, 125

[56] W. S. Massey, *Homology and cohomology theory: An approach based on Alexander-Spanier cochains*, Monographs and Textbooks in Pure and Applied Mathematics, Vol. 46, Marcel Dekker, Inc., New York-Basel, 1978. MR0488016 ↑86

[57] M. Matthey, *The Baum-Connes assembly map, delocalization and the Chern character*, Adv. Math. **183** (2004), no. 2, 316–379, DOI 10.1016/S0001-8708(03)00090-2. MR2041902 ↑117

[58] D. Meintrup and T. Schick, *A model for the universal space for proper actions of a hyperbolic group*, New York J. Math. **8** (2002), 1–7. MR1887695 ↑70, 72, 125

[59] R. Meyer, *Embeddings of derived categories of bornological modules*, Preprint, arXiv:math/0410596, 2004. ↑68, 73, 88, 95

[60] R. Meyer, *Combable groups have group cohomology of polynomial growth*, Q. J. Math. **57** (2006), no. 2, 241–261, DOI 10.1093/qmath/hai003. MR2237601 ↑68, 88, 95

[61] I. Mineyev and G. Yu, *The Baum-Connes conjecture for hyperbolic groups*, Invent. Math. **149** (2002), no. 1, 97–122, DOI 10.1007/s002220200214. MR1914618 ↑136

[62] H. Moscovici and F. Wu, *Index theory without symbols*, C^*-algebras: 1943–1993 (San Antonio, TX, 1993), Contemp. Math., vol. 167, Amer. Math. Soc., Providence, RI, 1994, pp. 304–351, DOI 10.1090/conm/167/1292020. MR1292020 ↑27, 54

[63] P. Piazza and T. Schick, *Bordism, rho-invariants and the Baum-Connes conjecture*, J. Noncommut. Geom. **1** (2007), no. 1, 27–111, DOI 10.4171/JNCG/2. MR2294190 ↑98, 132

[64] P. Piazza and T. Schick, *Groups with torsion, bordism and rho invariants*, Pacific J. Math. **232** (2007), no. 2, 355–378, DOI 10.2140/pjm.2007.232.355. MR2366359 ↑5, 112, 127, 128, 131

[65] P. Piazza and T. Schick, *Rho-classes, index theory and Stolz' positive scalar curvature sequence*, J. Topol. **7** (2014), no. 4, 965–1004, DOI 10.1112/jtopol/jtt048. MR3286895 ↑7, 10, 20, 60, 61, 102

[66] P. Piazza and T. Schick, *The surgery exact sequence, K-theory and the signature operator*, Ann. K-Theory **1** (2016), no. 2, 109–154, DOI 10.2140/akt.2016.1.109. MR3514938 ↑6, 7, 10

[67] P. Piazza and V. F. Zenobi, *Singular spaces, groupoids and metrics of positive scalar curvature*, J. Geom. Phys. **137** (2019), 87–123, DOI 10.1016/j.geomphys.2018.09.016. MR3893404 ↑61, 62

[68] M. Puschnigg, *New holomorphically closed subalgebras of C^*-algebras of hyperbolic groups*, Geom. Funct. Anal. **20** (2010), no. 1, 243–259, DOI 10.1007/s00039-010-0062-y. MR2647141 ↑9, 57, 67, 71, 72, 98

[69] T. Schick, *Real versus complex K-theory using Kasparov's bivariant KK-theory*, Algebr. Geom. Topol. **4** (2004), 333–346, DOI 10.2140/agt.2004.4.333. MR2077669 ↑117

[70] T. Schick, *The topology of positive scalar curvature*, Proceedings of the International Congress of Mathematicians—Seoul 2014. Vol. II, Kyung Moon Sa, Seoul, 2014, pp. 1285–1307. MR3728662 ↑6

[71] T. Schick and M. Seyedhosseini, *On an index theorem of Chang, Weinberger and Yu*, Münster J. Math. **14** (2021), no. 1, 123–154, DOI 10.17879/59019522628. MR4300165 ↑128

[72] T. Schick and V. F. Zenobi, *Positive scalar curvature due to the cokernel of the classifying map*, SIGMA Symmetry Integrability Geom. Methods Appl. **16** (2020), Paper No. 129, 12, DOI 10.3842/SIGMA.2020.129. MR4184617 ↑103, 126, 130, 132

[73] M. A. Shubin, *Pseudodifferential operators and spectral theory*, 2nd ed., Springer-Verlag, Berlin, 2001. Translated from the 1978 Russian original by Stig I. Andersson, DOI 10.1007/978-3-642-56579-3. MR1852334 ↑107

[74] Y. Song and X. Tang, *Higher orbit integrals, cyclic cocyles, and K-theory of reduced group C^*-algebra*, Preprint, `arXiv:1910.00175`, 2019. ↑139

[75] S. Stolz, *Concordance classes of positive scalar curvature metrics*, 1998. http://www3.nd.edu/ stolz/concordance.ps. ↑102, 114

[76] C. Wahl, *Higher ρ-invariants and the surgery structure set*, J. Topol. **6** (2013), no. 1, 154–192, DOI 10.1112/jtopol/jts028. MR3029424 ↑59

[77] C. Wahl, *The Atiyah-Patodi-Singer index theorem for Dirac operators over C^*-algebras*, Asian J. Math. **17** (2013), no. 2, 265–319, DOI 10.4310/AJM.2013.v17.n2.a2. MR3078932 ↑57, 59, 60

[78] C. A. Weibel, *An introduction to homological algebra*, Cambridge Studies in Advanced Mathematics, vol. 38, Cambridge University Press, Cambridge, 1994, DOI 10.1017/CBO9781139644136. MR1269324 ↑66, 67, 69

[79] S. Weinberger, Z. Xie, and G. Yu, *Additivity of higher rho invariants and nonrigidity of topological manifolds*, Comm. Pure Appl. Math. **74** (2021), no. 1, 3–113, DOI 10.1002/cpa.21962. MR4178180 ↑11, 135, 137, 138

[80] F. Wu, *The higher index theorem for manifolds with boundary*, J. Funct. Anal. **103** (1992), no. 1, 160–189, DOI 10.1016/0022-1236(92)90140-E. MR1144688 ↑94

[81] Z. Xie and G. Yu, *Higher rho invariants and the moduli space of positive scalar curvature metrics*, Adv. Math. **307** (2017), 1046–1069, DOI 10.1016/j.aim.2016.11.030. MR3590536 ↑116, 127

[82] Z. Xie and G. Yu, *Positive scalar curvature, higher rho invariants and localization algebras*, Adv. Math. **262** (2014), 823–866, DOI 10.1016/j.aim.2014.06.001. MR3228443 ↑7, 102

[83] Z. Xie and G. Yu, *Delocalized eta invariants, algebraicity, and K-theory of group C^*-algebras*, Int. Math. Res. Not. IMRN **15** (2021), 11731–11766, DOI 10.1093/imrn/rnz170. MR4294131 ↑13

[84] Z. Xie, G. Yu, and R. Zeidler, *On the range of the relative higher index and the higher rho-invariant for positive scalar curvature*, Adv. Math. **390** (2021), Paper No. 107897, 24, DOI 10.1016/j.aim.2021.107897. MR4292958 ↑11, 103, 116, 126, 127, 130, 131

[85] G. Yu, *Localization algebras and the coarse Baum-Connes conjecture*, K-Theory **11** (1997), no. 4, 307–318, DOI 10.1023/a:1007766031161. MR1451759 ↑7

[86] R. Zeidler, *Positive scalar curvature and product formulas for secondary index invariants*, J. Topol. **9** (2016), no. 3, 687–724, DOI 10.1112/jtopol/jtw005. MR3551834 ↑7, 102

[87] R. Zeidler, *Secondary large-scale index theory and positive scalar curvature*, Doctoral dissertation, Universität Göttingen, 2016. ↑117

[88] V. F. Zenobi, *Mapping the surgery exact sequence for topological manifolds to analysis*, J. Topol. Anal. **9** (2017), no. 2, 329–361, DOI 10.1142/S179352531750011X. MR3622237 ↑7, 102

[89] V. F. Zenobi, *The adiabatic groupoid and the Higson-Roe exact sequence*, J. Noncommut. Geom. **15** (2021), no. 3, 797–827, DOI 10.4171/jncg/422. MR4345202 ↑10, 15, 16, 20, 62, 63, 102, 105

[90] V. F. Zenobi, *Adiabatic groupoid and secondary invariants in K-theory*, Adv. Math. **347** (2019), 940–1001, DOI 10.1016/j.aim.2019.03.003. MR3922452 ↑7, 10, 32, 61, 62

Editorial Information

To be published in the *Memoirs*, a paper must be correct, new, nontrivial, and significant. Further, it must be well written and of interest to a substantial number of mathematicians. Piecemeal results, such as an inconclusive step toward an unproved major theorem or a minor variation on a known result, are in general not acceptable for publication.

Papers appearing in *Memoirs* are generally at least 80 and not more than 200 published pages in length. Papers less than 80 or more than 200 published pages require the approval of the Managing Editor of the Transactions/Memoirs Editorial Board. Published pages are the same size as those generated in the style files provided for $\mathcal{A}_{\mathcal{M}}\mathcal{S}$-LaTeX.

Information on the backlog for this journal can be found on the AMS website starting from `http://www.ams.org/memo`.

A Consent to Publish is required before we can begin processing your paper. After a paper is accepted for publication, the Providence office will send a Consent to Publish and Copyright Agreement to all authors of the paper. By submitting a paper to the *Memoirs*, authors certify that the results have not been submitted to nor are they under consideration for publication by another journal, conference proceedings, or similar publication.

Information for Authors

Memoirs is an author-prepared publication. Once formatted for print and on-line publication, articles will be published as is with the addition of AMS-prepared frontmatter and backmatter. Articles are not copyedited; however, confirmation copy will be sent to the authors.

Initial submission. The AMS uses Centralized Manuscript Processing for initial submissions. Authors should submit a PDF file using the Initial Manuscript Submission form found at `www.ams.org/submission/memo`, or send one copy of the manuscript to the following address: Centralized Manuscript Processing, MEMOIRS OF THE AMS, 201 Charles Street, Providence, RI 02904-2294 USA. If a paper copy is being forwarded to the AMS, indicate that it is for *Memoirs* and include the name of the corresponding author, contact information such as email address or mailing address, and the name of an appropriate Editor to review the paper (see the list of Editors below).

The paper must contain a *descriptive title* and an *abstract* that summarizes the article in language suitable for workers in the general field (algebra, analysis, etc.). The *descriptive title* should be short, but informative; useless or vague phrases such as "some remarks about" or "concerning" should be avoided. The *abstract* should be at least one complete sentence, and at most 300 words. Included with the footnotes to the paper should be the 2020 *Mathematics Subject Classification* representing the primary and secondary subjects of the article. The classifications are accessible from `www.ams.org/msc/`. The Mathematics Subject Classification footnote may be followed by a list of *key words and phrases* describing the subject matter of the article and taken from it. Journal abbreviations used in bibliographies are listed in the latest *Mathematical Reviews* annual index. The series abbreviations are also accessible from `www.ams.org/msnhtml/serials.pdf`. To help in preparing and verifying references, the AMS offers MR Lookup, a Reference Tool for Linking, at `www.ams.org/mrlookup/`.

Electronically prepared manuscripts. The AMS encourages electronically prepared manuscripts, with a strong preference for $\mathcal{A}_{\mathcal{M}}\mathcal{S}$-LaTeX. To this end, the Society has prepared $\mathcal{A}_{\mathcal{M}}\mathcal{S}$-LaTeX author packages for each AMS publication. Author packages include instructions for preparing electronic manuscripts, samples, and a style file that generates the particular design specifications of that publication series.

Authors may retrieve an author package for *Memoirs of the AMS* from `www.ams.org/journals/memo/memoauthorpac.html`. The *AMS Author Handbook* is available in PDF format from the author package link. The author package can also be obtained free of charge by sending email to `tech-support@ams.org` or from the Publication Division,

American Mathematical Society, 201 Charles St., Providence, RI 02904-2294, USA. When requesting an author package, please specify the publication in which your paper will appear. Please be sure to include your complete mailing address.

After acceptance. The source files for the final version of the electronic manuscript should be sent to the Providence office immediately after the paper has been accepted for publication. The author should also submit a PDF of the final version of the paper to the editor, who will forward a copy to the Providence office.

Accepted electronically prepared files can be submitted via the web at `www.ams.org/submit-book-journal/`, sent via FTP, or sent on CD to the Electronic Prepress Department, American Mathematical Society, 201 Charles Street, Providence, RI 02904-2294 USA. TeX source files and graphic files can be transferred over the Internet by FTP to the Internet node `ftp.ams.org` (130.44.1.100). When sending a manuscript electronically via CD, please be sure to include a message indicating that the paper is for the *Memoirs*.

Electronic graphics. Comprehensive instructions on preparing graphics are available at `www.ams.org/authors/journals.html`. A few of the major requirements are given here.

Submit files for graphics as EPS (Encapsulated PostScript) files. This includes graphics originated via a graphics application as well as scanned photographs or other computer-generated images. If this is not possible, TIFF files are acceptable as long as they can be opened in Adobe Photoshop or Illustrator.

Authors using graphics packages for the creation of electronic art should also avoid the use of any lines thinner than 0.5 points in width. Many graphics packages allow the user to specify a "hairline" for a very thin line. Hairlines often look acceptable when proofed on a typical laser printer. However, when produced on a high-resolution laser imagesetter, hairlines become nearly invisible and will be lost entirely in the final printing process.

Screens should be set to values between 15% and 85%. Screens which fall outside of this range are too light or too dark to print correctly. Variations of screens within a graphic should be no less than 10%.

Any graphics created in color will be rendered in grayscale for the printed version unless color printing is authorized by the Managing Editor and the Publisher. In general, color graphics will appear in color in the online version.

Inquiries. Any inquiries concerning a paper that has been accepted for publication should be sent to `memo-query@ams.org` or directly to the Electronic Prepress Department, American Mathematical Society, 201 Charles St., Providence, RI 02904-2294 USA.

Editors

The AMS uses Centralized Manuscript Processing for initial submissions to AMS journals. Authors should follow instructions listed on the Initial Submission page found at www.ams.org/tran/transubmit.html.

This journal is designed particularly for long research papers, normally at least 80 pages in length, and groups of cognate papers in pure and applied mathematics. Papers intended for publication in the *Memoirs* should be addressed to one of the following editors. The AMS uses Centralized Manuscript Processing for initial submissions to AMS journals. Authors should follow instructions listed on the Initial Submission page found at www.ams.org/memo/memosubmit.html

Managing Editor: Radha Kessar, Department of Mathematics, The Alan Turing Building, The University of Manchester, Oxford Road, Manchester M13 9PL United Kingdom; e-mail: radha.kessar@manchester.ac.uk

1. GEOMETRY, TOPOLOGY & LOGIC
Coordinating Editor: Mariya Ivanova Soskova, Department of Mathematics, University of Wisconsin–Madison, Madison, WI 53706 USA; e-mail: msoskova@math.wisc.edu

Algebraic topology, Bob Oliver, LAGA, Institut Galilée, Université Sorbonne Paris Nord, 99, Av. J-B Clément, 93430 Villetaneuse, France; e-mail: bobol@math.univ-paris13.fr

Arithmetic geometry and number theory, Megumi Harada, Department of Mathematics and Statistics, McMaster University, 1280 Main Street West, Hamilton, Ontario L8S 4K1, Canada; e-mail: haradam@mcmaster.ca

Geometric group theory and low-dimensional geometry and topology, Christopher Leininger, Department of Mathematics, Rice University, Houston, TX 77005 USA; e-mail: cjl12@rice.edu

Geometric topology, Jennifer Hom, School of Mathematics, Georgia Institute of Technology, 686 Cherry Street, Atlanta, GA 30332 USA; e-mail: jhom6@gatech.edu

Homotopy theory, higher category theory, and geometric applications, Clark Barwick, School of Mathematics, University of Edinburgh, James Clerk Maxwell Building, Peter Guthrie Tait Road, Edinburgh, EH9 3FD, United Kingdom; e-mail: clarkbarAMS@pm.me

Logic, Mariya Ivanova Soskova

2. ALGEBRA & NUMBER THEORY
Coordinating Editor: Daniel Krashen, 3E6 David Rittenhouse Lab., University of Pennsylvania, Department of Mathematics, 209 South 33rd Street, Philadelphia, PA 19104 USA; e-mail: daniel.krashen@gmail.com

Algebra, algebraic and arithmetic geometry, Daniel Krashen

Arithmetic algebraic geometry, Jennifer Balakrishnan, Department of Mathematics and Statistics, 665 Commonwealth Avenue, Boston University, Boston, MA 02215 USA; e-mail: jbala@bu.edu

Algebraic geometry, Renzo Cavalieri, Department of Mathematics, Colorado State University, Weber Building, Fort Collins, CO 80523 USA; e-mail: renzo@colostate.edu

Analytic number theory, Dimitris Koukoulopoulos, Université de Montréal, Pavillon André-Aisenstadt, 2920, chemin de la Tour, H3T 1J4, Montréal, QC, Canada; e-mail: dimitris.koukoulopoulos@umontreal.ca

Arithmetic geometry and number theory, David Loeffler, Faculty of Mathematics and Computer Science, UniDistance Suisse, Schinerstrasse 18, 3900 Brig-Glis, Switzerland; e-mail: david.loeffler@unidistance.ch

Associative rings and algebras, category theory, homological algebra, group theory and generalizations, Daniel Nakano, Department of Mathematics, University of Georgia, Athens, Georgia 30602 USA; e-mail: nakano@uga.edu

Commutative algebra, Henry Schenck, Department of Mathematics and Statistics, Auburn University, 315 Roosevelt Concourse, Auburn, AL 36849 USA; e-mail: hks0015@auburn.edu

Representation theory and group theory, Radha Kessar, Department of Mathematics, The Alan Turing Building, The University of Manchester, Oxford Road, Manchester M13 9PL United Kingdom; e-mail: radha.kessar@manchester.ac.uk

3. GEOMETRIC ANALYSIS & PDE
Coordinating Editor: Joachim Krieger, Bâtiment de Mathématiques, École Polytechnique Fédérale de Lausanne, Station 8, 1015 Lausanne Vaud, Switzerland; e-mail: joachim.krieger@epfl.ch

Elliptic and parabolic PDEs, geometric analysis, Yannick Sire, Department of Mathematics, Johns Hopkins University, 404 Krieger Hall, 3400 N. Charles Street, Baltimore, MD 21218 USA; e-mail: ysire1@jhu.edu

Elliptic and parabolic PDEs, geometric analysis, Ben Weinkove, Mathematics Department, Northwestern University, 2033 Sheridan Rd, Evanston, IL 60201 USA; e-mail: weinkove@math.northwestern.edu

Harmonic analysis and partial differential equations, Monica Visan, Department of Mathematics, University of California Los Angeles, 520 Portola Plaza, Los Angeles, CA 90095 USA; e-mail: visan@math.ucla.edu

Nonlinear Fourier and harmonic analysis and partial differential equations, Andrea R. Nahmod, Department of Mathematics and Statistics, University of Massachusetts Amherst, 710 N. Pleasant St. Lederle GRT, Amherst, MA 01003 USA; e-mail: nahmod@umass.edu

Partial differential equations, calculus of variations, optimizations and control theory, dynamical systems, Wilfrid Gangbo, College of Physical Sciences, UCLA, 520 Portola Plaza, Los Angeles, CA 90095 USA; e-mail: wgangbo@math.ucla.edu

Real analysis and partial differential equations, Joachim Krieger

4. ANALYSIS & DYNAMICS
Coordinating Editor: Krzysztof Frączek Faculty of Math and Computer Science, Nicolaus Copernicus University, Ul. Chopina 12/18 87-100 Toruń, Poland; e-mail: fraczek@mat.umk.pl

Functional analysis, probability and ergodic theory, Tim Austin, Department of Mathematics, University of California, Los Angeles, Los Angeles, CA 90095 USA; e-mail: tim@math.ucla.edu

Ergodic theory and dynamical systems, Krzysztof Frączek

Ergodic theory, applications to combinatorics and number theory, Nikos Frantzikinakis, University of Crete, Rethymno, Crete, Greece; e-mail: frantzikinakis@gmail.com

Functional analysis, groups and operator algebras, analysis in quantum information theory, Magdalena Musat, Department of Mathematical Sciences, University of Copenhagen, Universitetsparken 5, 2100 Copenhagen, Denmark; e-mail: musat@math.ku.dk

Operator algebras and ergodic theory, David Kerr, Mathematisches Institut, University of Münster, Einsteinstrasse 62, 48149 Münster, Germany; e-mail: kerrd@uni-muenster.de

5. PROBABILITY & COMBINATORICS
Coordinating Editor: Davar Khoshnevisan, Department of Mathematics, The University of Utah, Salt Lake City, UT 84112 USA; e-mail: davar@math.utah.edu

Combinatorics, Jacob Fox, Department of Mathematics, 450 Jane Stanford Way, Building 380, Stanford, CA 94305 USA; e-mail: jacobfox@stanford.edu

Enumeration, algebraic combinatorics, integrable probability, symmetric functions, Sylvie Corteel, Department of Mathematics, University of California, Berkeley, Berkeley, CA 94720 USA; CNRS, IRIF et Université Paris Cité, Paris, France; e-mail: corteel@berkeley.edu

Probability theory and statistical physics, Jian Ding, School of Mathematical Sciences, Peking University, Beijing 100871, China; e-mail: dingjian@math.pku.edu.cn

Probability theory and stochastic analysis, Davar Khoshnevisan

Probability theory, probabilistic combinatorics, statistical physics, Eyal Lubetzky, Department of Mathematics, Courant Institute of Mathematical Sciences, New York University, New York, NY 10012 USA; e-mail: eyal@courant.nyu.edu

SELECTED PUBLISHED TITLES IN THIS SERIES

1553 **Andreas Greven, Frank den Hollander, and Margriet Oomen,** Spatial Populations with Seed-Bank: Renormalisation on the Hierarchical Group, 2025

1552 **Jason Atnip, Hiroki Sumi, and Mariusz Urbański,** The Dynamics and Geometry of Semi-Hyperbolic Rational Semigroups, 2025

1551 **Ronan J. Conlon and Alix Deruelle,** Steady Gradient Kähler-Ricci Solitons on Crepant Resolutions of Calabi-Yau Cones, 2025

1550 **Jussi Behrndt, Fritz Gesztesy, and Marius Mitrea,** Sharp Boundary Trace Theory and Schrödinger Operators on Bounded Lipschitz Domains, 2025

1549 **Valentina Grazian and Chris Parker,** Saturated Fusion Systems on p-Groups of Maximal Class, 2025

1548 **Rui Loja Fernandes and Ivan Struchiner,** The Global Solutions to a Cartan's Realization Problem, 2025

1547 **Satoshi Kondo and Seidai Yasuda,** Arithmetic Quotients of the Bruhat-Tits Building for Projective General Linear Groups in Positive Characteristic, 2025

1546 **Filippo Mazzoli, Andrea Seppi, and Andrea Tamburelli,** Para-HyperKähler Geometry of the Deformation Space of Maximal Globally Hyperbolic Anti-de Sitter Three-Manifolds, 2025

1545 **Tobias Lenz,** G-Global Homotopy Theory and Algebraic K-Theory, 2025

1544 **James Maynard,** Primes in Arithmetic Progressions to Large Moduli III: Uniform Residue Classes, 2025

1543 **James Maynard,** Primes in Arithmetic Progressions to Large Moduli II: Well-Factorable Estimates, 2025

1542 **James Maynard,** Primes in Arithmetic Progressions to Large Moduli I: Fixed Residue Classes, 2025

1541 **Benedikt Ahrens, Paige Randall North, Michael Shulman, and Dimitris Tsementzis,** The Univalence Principle, 2025

1540 **J. Peter May, Mona Merling, and Angélica M. Osorno,** Equivariant Infinite Loop Space Theory: The Space Level Story, 2025

1539 **Rémi Coulon, Rhiannon Dougall, Barbara Schapira, and Samuel Tapie,** Twisted Patterson-Sullivan Measures and Applications to Amenability and Coverings, 2025

1538 **Jeffrey S. Case,** The Bigraded Rumin Complex via Differential Forms, 2025

1537 **Michael Drmota, Clemens Müllner, and Lukas Spiegelhofer,** Primes as Sums of Fibonacci Numbers, 2025

1536 **Matthew Baker and Oliver Lorscheid,** Foundations of Matroids, Part 1: Matroids without Large Uniform Minors, 2025

1535 **Christian Stump, Hugh Thomas, and Nathan Williams,** Cataland: Why the Fuß?, 2025

1534 **Padraig Condon, Alberto Espuny Díaz, António Girão, Daniela Kühn, and Deryk Osthus,** Hamiltonicity of Random Subgraphs of the Hypercube, 2024

1533 **Peter Rudzis,** Rough Collisions, 2024

1532 **Xiang Yu,** On the Problem of Infinite Spin in Total Collisions of the Planar N-Body Problem, 2024

1531 **Dirk Basson, Florian Breuer, and Richard Pink,** Drinfeld Modular Forms of Arbitrary Rank, 2024

1530 **Jianbei An, Gerhard Hiss, and Frank Lübeck,** The Inductive Blockwise Alperin Weight Condition for the Chevalley Groups $F_4(q)$, 2024

For a complete list of titles in this series, visit the
AMS Bookstore at **www.ams.org/bookstore/memoseries/**.